Lecture Notes in Physics

The Lecture Notes in Physics

The series Lecture Notes in Physics (LNP), founded in 1969, reports new developments in physics research and teaching – quickly and informally, but with a high quality and the explicit aim to summarize and communicate current knowledge in an accessible way. Books published in this series are conceived as bridging material between advanced graduate textbooks and the forefront of research and to serve three purposes:

- to be a compact and modern up-to-date source of reference on a well-defined topic

- to serve as an accessible introduction to the field to postgraduate students and nonspecialist researchers from related areas

- to be a source of advanced teaching material for specialized seminars, courses and schools

Both monographs and multi-author volumes will be considered for publication. Edited volumes should, however, consist of a very limited number of contributions only. Proceedings will not be considered for LNP.

Volumes published in LNP are disseminated both in print and in electronic formats, the electronic archive being available at springerlink.com. The series content is indexed, abstracted and referenced by many abstracting and information services, bibliographic networks, subscription agencies, library networks, and consortia.

Proposals should be sent to a member of the Editorial Board, or directly to the managing editor at Springer:

Christian Caron
Springer Heidelberg
Physics Editorial Department I
Tiergartenstrasse 17
69121 Heidelberg / Germany
christian.caron@springer.com

S. Massaglia
G. Bodo
A. Mignone
P. Rossi (Eds.)

Jets From Young Stars III

Numerical MHD and Instabilities

 Springer

Editors

Silvano Massaglia
Università di Torino
Dipto. Fisica Generale
Via Pietro Giuria, 1
10125 Torino
Italy
massaglia@ph.unito.it

Gianluigi Bodo
Osservatorio Astronomico di
Torino
Strada Osservatorio, 20
10025 Pino Torinese TO
Italy
bodo@to.astro.it

Andrea Mignone
Università di Torino
Dipto. Fisica Generale
Via Pietro Giuria, 1
10125 Torino
Italy
mignone@ph.unito.it

Paola Rossi
INAF Osservatorio
di Torino Astronomico
Viale Osservatorio, 20
10025 Pino Torinese
Italy
rossi@to.astro.it

Massaglia, S. et al. (Eds.), *Jets From Young Stars III*, Lect. Notes Phys. 754 (Springer, Berlin Heidelberg 2008), DOI 10.1007/ 978 3 540-76967-5

ISBN: 978-3-540-76966-8   e-ISBN: 978-3-540-76967-5

DOI 10.1007/978-3-540-76967-5

Lecture Notes in Physics ISSN: 0075-8450   e-ISSN: 1616-6361

Library of Congress Control Number: 2008927516

Cover design: eStudio Calamar S.L., F. Steinen-Broo, Pau/Girona, Spain

Printed on acid-free paper

9 8 7 6 5 4 3 2 1

springer.com

Preface

The study of the mechanisms that govern origin and propagation of stellar jets involves the treatment of many concurrent physical processes such as gravitation, hydrodynamics and magnetohydrodynamics, atomic physics and radiation. In the past years, an intensive work has been done looking for solutions of the ideal MHD equations in the steady state limit as well as studying the stability of outflows in the linear regime. These kind, of approaches have provided a contribution to the understanding of jets that can hardly be overestimated. However, the extension of the analyses to the time-dependent and nonlinear regimes could not be avoided, and the MHD numerical simulations were the only mean to achieve this goal.

In the recent years, considerable progresses have been made by the computational fluid dynamic community in the development of numerical techniques, the so-called high resolution shock capturing schemes, well suited for the treatment of supersonic flows with discontinuities. The numerical simulations of astrophysical jets took advantage of these developments; however new physics needed to be incorporated, such as magnetic field effects, radiation losses by diluted gases, and proper astrophysics environments. These needs led to the nontrivial extension of the methods devised for the Euler equations of gasdynamics to the magneto-hydrodynamical system. On the other hand, the possibility of carrying out numerical calculations has been greatly facilitated by the availability, on one hand, of powerful supercomputers and, on the other hand, of fast processors at low cost. Large scale 3D simulations of jets at high resolution are now possible thanks to supercomputers, but also high resolution 2D MHD simulations can be performed routinely on desktop computers. These possibilities have greatly extended our understanding of jets, and numerical simulations are now an essential tool for investigating the physics of such objects.

This book collects the lectures from the third JETSET school, Jets from Young Stars III: Numerical MHD and Instabilities, held in Sauze d'Oulx in January 2007. The aim of this school was to introduce PhD students and young researchers to the basic methods in computational hydrodynamics and magneto-hydrodynamics as well as to review some of the most relevant insta-bilities in astrophysical outflows.

The book is divided in two parts. In the first one, Eleuterio Toro presents and discusses the basic numerical methods in hydrodynamics, their stabil-ity, monotonicity and accuracy properties, and introduces different Riemann solvers and flux limiter methods. Andrea Mignone and Gianluigi Bodo then extend the discussion to MHD, examining, in particular, the various methods devised to overcome the numerical difficulties intrinsic to the MHD equations. The second part is devoted to hydrodynamic and MHD instabilities, such as the Kelvin–Helmholtz instabilities, treated by Edoardo Trussoni, the pressure driven instabilities, by Pierre-Yves Longaretti, the thermal instabilities, by Gianluigi Bodo, and the instabilities in radiative shocks, by Andrea Mignone.

The editors are grateful to the lecturers for their presentations and efforts to contribute to this book, and to Ovidiu Tesileanu and Titos Matsakos that have been of great help in the organization of the school. Thanks also to Eileen Flood and Gabriella Ardizzoia for their precious help in managing the life of the participants during the days of the school. The editors acknowledge the invaluable contribution and support by Emma Whelan for taking care of the book editing.

Torino   *Silvano Massaglia*
                                                          Andrea Mignone
Pino Torinese   *Gianluigi Bodo*
                                                          Paola Rossi

Contents

Part I Numerical Methods

Computational Methods for Hyperbolic Equations

E.F. Toro 3
1   Equations 4
2   The Finite Difference Method 16
3   Two Riemann Solvers: HLLC and EVILIN 31
4   Non-linear Methods for Scalar Equations 43
5   Non-linear Schemes for Hyperbolic Systems 55
References 67

Shock-Capturing Schemes in Computational MHD

A. Mignone and G. Bodo 71
1   Introduction 71
2   The MHD Equations 72
3   The Riemann Problem in MHD 73
4   The $\nabla \cdot B = 0$ Condition 91
References 99

Part II Hydrodynamic and Magneto-Hydrodynamic Instabilities

The Kelvin–Helmholtz Instability

E. Trussoni 105
1   Introduction 106
2   Definitions 106
3   Equations and Mathematical Approach 107
4   KHI: Linear Analysis 109
5   KHI: Nonlinear Evolution 121
6   Conclusions 126
References 129

Pressure-Driven Instabilities in Astrophysical Jets

P.-Y. Longaretti 131
1   Introduction 132
2   Heuristic Description of MHD Instabilities 134
3   Ideal MHD in Static Columns 137
4   The Energy Principle and Its Consequences 140
5   Dispersion Relation in the Large Azimuthal Field Limit 142
6   Moving Columns 146
7   Summary and Open Issues 147
References 150

Thermal Instabilities

G. Bodo 153
1   Introduction 154
2   Physical Discussion 154
3   Linear Analysis 156
4   Influence of Radiative Losses on the KH Instability 161
References 162

The Oscillatory Instability of Radiative Shock Waves

A. Mignone 163
1   Introduction 164
2   Linear Theory 165
3   Nonlinear Dynamics 171
4   Discussion 174
References 175

Index 177

Part I

Numerical Methods

Computational Methods for Hyperbolic Equations

E.F. Toro

Laboratory of Applied Mathematics, Department of Civil and Environmental
Engineering, University of Trento, Trento, Italy,
toro@ing.unitn.it

Abstract. This is an introduction to some of the basic concepts on modern numerical methods for computing approximate solutions to hyperbolic partial differential equations. This chapter is divided into five sections. Section 1 contains a review of some elementary theoretical concepts on hyperbolic equations, mainly focused on the linear case; the Riemann problem for a general linear system with constant coefficients is solved in detail. Section 2 is an introduction to the basics of discretization methods, including finite difference methods and finite volume methods; concepts such as local truncation error, linear stability and modified equation are included; Godunov's theorem is stated, proved and its implications are discussed. Section 3 contains two approximate Riemann solvers, as applied to the three-dimensional Euler equations, namely HLLC and EVILIN. Section 4 deals with the construction of non-linear (non-oscillatory) numerical methods of the TVD and ENO type, for a scalar conservation law. In Sect. 5 we use the theory developed for scalar equations as

Toro, E.F.: *Computational Methods for Hyperbolic Equations.* Lect. Notes Phys. **754**, 3–69
(2008)
DOI 10.1007/978-3-540-76967-5_1

a guideline to construct non-linear (quasi non-oscillatory) second-order finite volume schemes for one-dimensional non-linear systems with source terms. Key references for further reading are indicated at the end of each section.

Keywords Hyperbolic Equations · Riemann Problem · Godunov Methods · Riemann Solvers · Non-linear Schemes · Source Terms

1 Equations

Here we study some very basic properties of hyperbolic equations, starting from the general setting of hyperbolic balance laws in three space dimensions. The linear advection equation with constant coefficient and associated characteristic curves are studied in detail. The general initial value problem for general linear hyperbolic systems is solved, as is the Riemann problem. A few useful notes on non-linear equations are made. In particular we introduce the integral form of the conservation laws. This is useful for at least two reasons. First, hyperbolic equations admit discontinuous solutions even if the initial data is smooth. Shock waves are formed in finite time. The differential form of the equations is no longer valid, while the integral form is and accommodates an extension of the set of admissible solutions. However the extended set of solutions is too large and includes solutions that have no physical value. Entropy criteria are enforced to select the physically admissible solutions. Then, the integral form also allows quite directly the construction of numerical methods of the finite volume type. References for further study are given at the end of the section. We study some basic concepts on hyperbolic balance laws

$$\partial_t \mathbf{Q} + \partial_x \mathbf{F}(\mathbf{Q}) + \partial_y \mathbf{G}(\mathbf{Q}) + \partial_z \mathbf{H}(\mathbf{Q}) = \mathbf{S}(\mathbf{Q}), \tag{1}$$

where

$$\mathbf{Q} = \begin{bmatrix} q_1 \\ q_2 \\ \dots \\ q_m \end{bmatrix}, \mathbf{F} = \begin{bmatrix} f_1 \\ f_2 \\ \dots \\ f_m \end{bmatrix}, \mathbf{G} = \begin{bmatrix} g_1 \\ g_2 \\ \dots \\ g_m \end{bmatrix}, \mathbf{H} = \begin{bmatrix} h_1 \\ h_2 \\ \dots \\ h_m \end{bmatrix}, \mathbf{S} = \begin{bmatrix} s_1 \\ s_2 \\ \dots \\ s_m \end{bmatrix}. \tag{2}$$

The independent variables are x, y, z and t. $\mathbf{Q}$ is the vector of dependent variables, called *conserved variables*, and are the unknowns of the problem; $\mathbf{F}(\mathbf{Q})$ is the flux vector in the x-direction; $\mathbf{G}(\mathbf{Q})$ is the flux vector in the y-direction; $\mathbf{H}(\mathbf{Q})$ is the flux vector in the z-direction; and $\mathbf{S}(\mathbf{Q})$ is the vector of source terms, which are prescribed functions of the unknown $\mathbf{Q}$ and no differential terms are involved here. We use $\partial_d \mathbf{X}$ to denote the first-order partial derivative of any vector $\mathbf{X}$ with respect to d, for which we also use $\mathbf{X}_d$. Note that for each component i of the vectors in (2) we have

$$\left.\begin{aligned}
q_i &= q_i(x, y, z, t),\\
f_i &= f_i(q_1(x, y, z, t), \ldots, q_m(x, y, z, t)),\\
g_i &= g_i(q_1(x, y, z, t), \ldots, q_m(x, y, z, t)),\\
h_i &= h_i(q_1(x, y, z, t), \ldots, q_m(x, y, z, t)),\\
s_i &= s_i(x, y, z, t, q_1(x, y, z, t), \ldots, q_m(x, y, z, t)).
\end{aligned}\right\}$$

A system like (1) is said to be a system of conservation laws (with source terms) written in *differential conservative form*. This form is valid only for smooth solutions. For solutions with discontinuities, one resorts to the integral form of the equations, as seen later.

Scalar One-dimensional Examples. A scalar (a single equation) conservation law with source term in one space dimension reads

$$\partial_t q + \partial_x f(q) = s(q), \tag{3}$$

where $q(x, t)$ is the conserved variable, $f(q)$ is the flux function and $s(q)$ is the source term. An equation of the form (3) is said to be *homogeneous* if the source term is identically zero; otherwise it is called *inhomogeneous*. In order to have a determined equation, we need to define the flux and the source term, as illustrated in the examples below:

- The (homogeneous) linear advection equation

$$\partial_t q + \lambda \partial_x q = 0, \quad f(q) = \lambda q \ (\lambda, \text{ a constant}), \quad s(q) = 0. \tag{4}$$

- The (inhomogeneous) linear advection equation with a source term

$$\partial_t q + \lambda \partial_x q = \beta q, \quad f(q) = \lambda q \ (\lambda, \text{ a constant}), \quad s(q) = \beta q \ (\beta, \text{ a constant}). \tag{5}$$

- The inviscid Burgers equation

$$\partial_t q + \partial_x \left(\frac{1}{2} q^2\right) = 0, \quad f(q) = \frac{1}{2} q^2, \quad s(q) = 0. \tag{6}$$

- The traffic flow equation

$$\partial_t q + \partial_x \left(u_{max}(1 - q/q_{max})q\right) = 0, \quad f(q) = u_{max}(1 - q/q_{max})q, \quad s(q) = 0, \tag{7}$$

where u_{max} and q_{max} are constants.

A System: The Isentropic Euler Equations. The equations in one space dimension are

$$\partial_t \mathbf{Q} + \partial_x \mathbf{F}(\mathbf{Q}) = \mathbf{0}, \tag{8}$$

with

$$\mathbf{Q} = \begin{bmatrix} q_1 \\ q_2 \end{bmatrix} \equiv \begin{bmatrix} \rho \\ \rho u \end{bmatrix}, \quad \mathbf{F}(\mathbf{Q}) = \begin{bmatrix} f_1 \\ f_2 \end{bmatrix} \equiv \begin{bmatrix} \rho u \\ \rho u^2 + p(\rho) \end{bmatrix}. \tag{9}$$

Here $\rho(x,t)$ is density, $u(x,t)$ is particle velocity and $p(\rho)$ is pressure. As there are more unknowns than equations, we need to specify a closure condition, which is generally done by prescribing the pressure p as a function of density, $p = p(\rho)$. This closure condition is also named *equation of state* (EOS); the example considered is an isentropic EOS.

1.1 The Linear Advection Equation and Characteristics

We consider various forms of scalar linear advection partial differential equation (PDE) (4) and associated characteristic curves.

Linear Advection with Constant Coefficient

Let us first consider the *model* homogeneous hyperbolic equation

$$\partial_t q + \lambda \partial_x q = 0, \quad -\infty < x < \infty, \ t \geq 0 \tag{10}$$

and curves $x = x(t)$ in the x–t half-plane that are associated to the initial-value problem (IVP) for an ordinary differential equation (ODE) with an initial condition (IC), namely

$$\left.\begin{array}{ll} \text{ODE}: & \dfrac{dx}{dt} = \lambda, \\[2mm] \text{IC} & : x(0) = x_0, \end{array}\right\} \tag{11}$$

where x_0, called the *foot of the characteristic*, is the intersection of the curve $x(t)$ with the x-axis at $t = 0$. Solutions of (11) are called *characteristic curves* and have the form

$$x = x_0 + \lambda t, \tag{12}$$

which are straight lines. The time rate of change of $q(x(t),t)$ along a characteristic curve $x = x(t)$ is

$$\frac{dq}{dt} = \frac{\partial q}{\partial t}\frac{dt}{dt} + \frac{\partial q}{\partial x}\frac{dx}{dt} = \partial_t q + \lambda \partial_x q = 0. \tag{13}$$

Consequently, the characteristic curves are curves $x(t)$ such that:

- the PDE becomes an ODE along $x(t)$, namely $dq/dt = 0$ and
- the ODE states that the unknown $q(x,t)$ is constant along the characteristic curve.

We now consider the general IVP for the linear advection equation

$$\left.\begin{array}{ll} \text{PDE}: \partial_t q + \lambda \partial_x q = 0, & -\infty < x < \infty, t > 0, \\[2mm] \text{IC} \quad : q(x,0) = q^{(0)}(x), & -\infty < x < \infty, \end{array}\right\} \tag{14}$$

where $q^{(0)}(x)$ is the initial condition for the PDE, a prescribed function of distance x.

The solution of IVP (14) at any point (x,t) is found by considering the characteristic curve that passes through the point (x,t) and its foot x_0. As the solution $q(x,t)$ is a constant along the characteristic curve, the $q(x,t)$ is the same as the initial condition $q(x,0) = q^{(0)}(x)$ at the initial point x_0, that is $q(x,t) = q^{(0)}(x_0)$. Using (12) we get $x_0 = x - \lambda t$ and thus the solution can be written as

$$q(x,t) = q^{(0)}(x_0) = q^{(0)}(x - \lambda t). \tag{15}$$

Exercise. Verify that the given function $q(x,t)$ is a solution of the problem.

The Riemann Problem

The Riemann problem is the special IVP

$$\left.\begin{array}{l} \text{PDE} : \partial_t q + \lambda \partial_x q = 0, \quad -\infty < x < \infty, \ t > 0 \\[2mm] \text{IC} : q(x,0) = q^{(0)}(x) = \left\{\begin{array}{l} q_{\mathrm{L}} \text{ (constant) if } x < 0, \\[2mm] q_{\mathrm{R}} \text{(constant) if } x > 0. \end{array}\right. \end{array}\right\} \tag{16}$$

From (15) the solution is

$$q(x,t) = q^{(0)}(x - \lambda t) = \left\{\begin{array}{l} q_{\mathrm{L}} \text{ if } x - \lambda t < 0 \Leftrightarrow \dfrac{x}{t} < \lambda, \\[3mm] q_{\mathrm{R}} \text{ if } x - \lambda t > 0 \Leftrightarrow \dfrac{x}{t} > \lambda. \end{array}\right. \tag{17}$$

That is, the solution of the Riemann problem is

$$q(x,t) = \left\{\begin{array}{l} q_{\mathrm{L}} \text{ if } \dfrac{x}{t} < \lambda, \\[3mm] q_{\mathrm{R}} \text{ if } \dfrac{x}{t} > \lambda. \end{array}\right. \tag{18}$$

Linear Advection with Variable Coefficient

We now consider the general IVP for the linear advection equation with variable coefficient $\lambda(x,t)$, namely

$$\left.\begin{array}{ll} \text{PDE} : \partial_t q + \lambda(x,t)\partial_x q = 0, & -\infty < x < \infty, \ t > 0, \\[2mm] \text{IC} \quad : q(x,0) = q^{(0)}(x), & -\infty < x < \infty, \end{array}\right\} \tag{19}$$

where again $q^{(0)}(x)$ is a prescribed function of distance x. The characteristics curves are the solutions of

$$\left.\begin{array}{ll} \text{ODE} : \dfrac{dx}{dt} = \lambda(x,t), \\[3mm] \text{IC} \quad : x(0) = x_0 \end{array}\right\} \tag{20}$$

and the curves are no longer straight lines, in general.

Linear Advection with a Linear Source Term

We now consider the general IVP for the linear advection equation with constant coefficient λ and a linear source term

$$\left.\begin{array}{l} \text{PDE} : \partial_t q + \lambda \partial_x q = \beta q, \quad -\infty < x < \infty, \; t > 0, \\[2mm] \text{IC} : q(x,0) = q^{(0)}(x), \quad -\infty < x < \infty, \end{array}\right\} \tag{21}$$

where again $q^{(0)}(x)$ is a prescribed function of distance x, and λ and β are constants. It is easily seen (verify) that the exact solution is

$$q(x,t) = q^{(0)}(x - \lambda t)e^{\beta t}. \tag{22}$$

1.2 Quasi-linear Form and Hyperbolicity

Consider the general non-linear homogeneous scalar conservation law (3) and apply the chain rule to the flux term. We obtain

$$\partial_t q + \partial_x f(q) = \partial_t q + \partial_q f(q)\partial_x q = 0, \tag{23}$$

or

$$\partial_t q + \lambda(q)\partial_x q = 0, \quad \lambda(q) = \partial_q f(q) = \frac{\partial f(q)}{\partial q} \equiv f'(q). \tag{24}$$

This is the *quasi-linear form* of the equation and the coefficient $\lambda(q)$ is the characteristic speed, a function of the unknown q.

Also a system can be expressed in quasi-linear form. To illustrate this, we consider the following non-linear system of two equations

$$\partial_t \mathbf{Q} + \partial_x \mathbf{F}(\mathbf{Q}) = \mathbf{0}, \tag{25}$$

where

$$\mathbf{Q} = \begin{bmatrix} q_1 \\ q_2 \end{bmatrix}, \quad \mathbf{F}(\mathbf{Q}) = \begin{bmatrix} f_1(q_1, q_2) \\ f_2(q_1, q_2) \end{bmatrix}. \tag{26}$$

Applying the chain rule to the flux term for each equation, we have

$$\left.\begin{array}{l} \partial_t q_1 + \partial_{q_1} f_1(q_1, q_2)\partial_x q_1 + \partial_{q_2} f_1(q_1, q_2)\partial_x q_2 = 0, \\[2mm] \partial_t q_2 + \partial_{q_1} f_2(q_1, q_2)\partial_x q_1 + \partial_{q_2} f_2(q_1, q_2)\partial_x q_2 = 0. \end{array}\right\}$$

In matrix form, the system reads

$$\partial_t \begin{bmatrix} q_1 \\ q_2 \end{bmatrix} + \begin{bmatrix} \frac{\partial f_1}{\partial q_1} & \frac{\partial f_1}{\partial q_2} \\[2mm] \frac{\partial f_2}{\partial q_1} & \frac{\partial f_2}{\partial q_2} \end{bmatrix} \partial_x \begin{bmatrix} q_1 \\ q_2 \end{bmatrix} = \begin{bmatrix} 0 \\ 0 \end{bmatrix}, \tag{27}$$

or

$$\partial_t \mathbf{Q} + \mathbf{A}(\mathbf{Q})\partial_x \mathbf{Q} = \mathbf{0}, \tag{28}$$

where

$$\mathbf{A}(\mathbf{Q}) = \frac{\partial \mathbf{F}}{\partial \mathbf{Q}} = \begin{bmatrix} \frac{\partial f_1}{\partial q_1} & \frac{\partial f_1}{\partial q_2} \\ \\ \frac{\partial f_2}{\partial q_1} & \frac{\partial f_2}{\partial q_2} \end{bmatrix} \tag{29}$$

is called the *Jacobian matrix*. For an $m \times m$ one-dimensional non-linear system of the form (25), the quasi-linear form is like (28) with Jacobian matrix

$$\mathbf{A}(\mathbf{Q}) = \frac{\partial \mathbf{F}}{\partial \mathbf{Q}} = \begin{bmatrix} \frac{\partial f_1}{\partial q_1} & \frac{\partial f_1}{\partial q_2} & \cdots & \frac{\partial f_1}{\partial q_m} \\ \\ \frac{\partial f_2}{\partial q_1} & \frac{\partial f_2}{\partial q_2} & \cdots & \frac{\partial f_2}{\partial q_m} \\ \cdots & \cdots & \cdots & \cdots \\ \frac{\partial f_m}{\partial q_1} & \frac{\partial f_m}{\partial q_2} & \cdots & \frac{\partial f_m}{\partial q_m} \end{bmatrix}. \tag{30}$$

Eigenvalues and eigenvectors. Recall that the eigenvalues of a matrix $\mathbf{A}$ are the roots of the characteristic polynomial

$$|\mathbf{A} - \lambda \mathbf{I}| = 0, \tag{31}$$

where $\mathbf{I}$ is the identity matrix and λ is a parameter. A right eigenvector of $\mathbf{A}$ corresponding to an eigenvalue λ is a vector $\mathbf{R}$ such that $\mathbf{A}\mathbf{R} = \lambda\mathbf{R}$. Similarly, a left eigenvector of $\mathbf{A}$ corresponding to an eigenvalue λ is a vector $\mathbf{L}$ such that $\mathbf{L}\mathbf{A} = \lambda\mathbf{L}$.

Hyperbolic system. A one-dimensional $m \times m$ first-order system of the form (25) is hyperbolic if the Jacobian matrix $\mathbf{A}(\mathbf{Q})$ in (30) has m real eigenvalues $\lambda_1, \lambda_2, \ldots, \lambda_m$ and a corresponding set of m linearly independent eigenvectors $\mathbf{R}_1, \mathbf{R}_2, \ldots, \mathbf{R}_m$. The system is said to be *strictly hyperbolic* if it is hyperbolic and all eigenvalues are distinct.

Three-dimensional hyperbolic system. A three-dimensional $m \times m$ first-order system of the form (1) is hyperbolic if the matrix

$$\mathbf{D}(\mathbf{Q}) = \omega_1 \mathbf{A}(\mathbf{Q}) + \omega_2 \mathbf{B}(\mathbf{Q}) + \omega_3 \mathbf{C}(\mathbf{Q}) \tag{32}$$

has m real eigenvalues $\lambda_1, \lambda_2, \ldots, \lambda_m$ and a corresponding set of m linearly independent eigenvectors $\mathbf{R}_1, \mathbf{R}_2, \ldots, \mathbf{R}_m$, for all linear combinations (32), where the coefficients ω_1, ω_2 and ω_3 define a non-zero vector, that is

$$\sqrt{\omega_1^2 + \omega_2^2 + \omega_3^2} > 0.$$

$\mathbf{A}$, $\mathbf{B}$ and $\mathbf{C}$ are the Jacobian matrices corresponding to the fluxes $\mathbf{F}$, $\mathbf{G}$ and $\mathbf{H}$ in (1).

Example. **Eigenstructure of the Isentropic Equations.** A non-linear example of a system of conservation laws are the isentropic equations of gas dynamics (8) and (9), together with an isentropic EOS,

$$p = C\rho^\gamma, \quad C = \text{constant}, \quad \gamma = \text{constant}. \tag{33}$$

The reader is invited to calculate the Jacobian matrix, the eigenvalues and the right eigenvectors and to show that for a generalized isentropic EOS, $p = p(\rho)$, the system is hyperbolic if and only if $p'(\rho) > 0$, that is, the pressure must be a monotone increasing function of ρ. Show also that the sound speed has the general form

$$a = \sqrt{p'(\rho)}. \tag{34}$$

The eigenvalues are

$$\lambda_1 = u - a, \quad \lambda_2 = u + a, \tag{35}$$

and the right eigenvectors are

$$\mathbf{R}_1 = \begin{bmatrix} 1 \\ u - a \end{bmatrix}, \quad \mathbf{R}_2 = \begin{bmatrix} 1 \\ u + a \end{bmatrix}, \tag{36}$$

with the sound speed a as claimed.

1.3 Linear Systems

In the previous section, we studied in detail the behaviour and the general solution of the simplest PDE of hyperbolic type, namely the linear advection equation with constant wave propagation speed. Here we extend the analysis to sets of m hyperbolic PDEs of the form

$$\partial_t \mathbf{Q} + \mathbf{A} \partial_x \mathbf{Q} = \mathbf{0}, \tag{37}$$

where the coefficient matrix $\mathbf{A}$ is constant. From the assumption of hyperbolicity, $\mathbf{A}$ has m real eigenvalues λ_i and m linearly independent eigenvectors $\mathbf{R}_i$, $i = 1, \ldots, m$.

Diagonalisation and Characteristic Variables

In order to analyse and solve the general IVP for (37), it is found useful to transform the dependent variables $\mathbf{Q}(x, t)$ to a new set of dependent variables $\mathbf{W}(x, t)$. To this end we recall the following definition.

Diagonalisable System. A matrix $\mathbf{A}$ is said to be diagonalisable if it can be expressed as

$$\mathbf{A} = \mathbf{R}\Lambda\mathbf{R}^{-1} \text{ or } \Lambda = \mathbf{R}^{-1}\mathbf{A}\mathbf{R}, \tag{38}$$

in terms of a diagonal matrix Λ and a matrix $\mathbf{R}$. The diagonal elements of Λ are the eigenvalues λ_i of $\mathbf{A}$ and the columns $\mathbf{R}_i$ of $\mathbf{R}$ are the right eigenvectors of $\mathbf{A}$ corresponding to the eigenvalues λ_i, that is

$$
\Lambda = \begin{bmatrix} \lambda_1 & \cdots & 0 \\ 0 & \cdots & 0 \\ \vdots & \vdots & \vdots \\ 0 & \cdots & \lambda_m \end{bmatrix}, \quad \mathbf{R} = [\mathbf{R}_1, \ldots, \mathbf{R}_m], \quad \mathbf{AR}_i = \lambda_i \mathbf{R}_i. \tag{39}
$$

A system (37) is said to be *diagonalisable* if the coefficient matrix $\mathbf{A}$ is diagonalisable. Based on the concept of diagonalisation, one often defines a hyperbolic system (37) as a system with real eigenvalues and diagonalisable coefficient matrix.

Characteristic Variables

The existence of the inverse matrix $\mathbf{R}^{-1}$ makes it possible to define a new set of dependent variables $\mathbf{W} = [w_1, w_2, \ldots, w_m]^{\mathrm{T}}$ via the transformation

$$
\mathbf{W} = \mathbf{R}^{-1}\mathbf{Q} \text{ or } \mathbf{Q} = \mathbf{RW}, \tag{40}
$$

so that the linear system (37), when expressed in terms of $\mathbf{W}$, becomes *completely decoupled* in a sense to be defined. The new variables $\mathbf{W}$ are called *characteristic variables*. Next we derive the governing PDEs in terms of the characteristic variables, for which we need the partial derivatives in equation (37). As $\mathbf{A}$ is a constant, $\mathbf{R}$ is a also constant and, therefore, these derivatives are

$$
\partial_t \mathbf{Q} = \mathbf{R}\partial_t \mathbf{W}, \quad \partial_x \mathbf{Q} = \mathbf{R}\partial_x \mathbf{W}.
$$

Direct substitution of these expressions into (37) gives

$$
\mathbf{R}\partial_t \mathbf{W} + \mathbf{AR}\partial_x \mathbf{W} = \mathbf{0}.
$$

Multiplication of this equation from the left by $\mathbf{R}^{-1}$ and use of (40) gives

$$
\partial_t \mathbf{W} + \Lambda \partial_x \mathbf{W} = \mathbf{0}. \tag{41}
$$

This is called the *canonical form* or *characteristic form* of system (37). When written in full this system becomes

$$
\begin{bmatrix} w_1 \\ w_2 \\ \vdots \\ w_m \end{bmatrix}_t + \begin{bmatrix} \lambda_1 & \cdots & 0 \\ 0 & \cdots & 0 \\ \vdots & \vdots & \vdots \\ 0 & \cdots & \lambda_m \end{bmatrix} \begin{bmatrix} w_1 \\ w_2 \\ \vdots \\ w_m \end{bmatrix}_x = \mathbf{0}. \tag{42}
$$

Clearly the ith PDE of this system is

$$
\frac{\partial w_i}{\partial t} + \lambda_i \frac{\partial w_i}{\partial x} = 0, \quad i = 1, \ldots, m \tag{43}
$$

and involves the *single unknown* $w_i(x, t)$; the system is, therefore, *decoupled*; equation (43) is identical to the linear advection equation (4); now the characteristic speed is λ_i and there are m characteristic curves satisfying m ODEs

$$
\frac{dx}{dt} = \lambda_i, \quad \text{for } i = 1, \ldots, m.
$$

The General Initial-Value Problem

We now study the general IVP for a linear hyperbolic system

$$\text{PDEs}: \partial_t \mathbf{Q} + \mathbf{A}\partial_x \mathbf{Q} = \mathbf{0}, \quad -\infty < x < \infty, t > 0, \left.\begin{array}{c} \\ \\ \end{array}\right\}$$

$$\text{IC} \quad : \quad \mathbf{Q}(x,0) = \mathbf{Q}^{(0)}(x) = \left\{ \begin{array}{ll} \mathbf{Q_L} & \text{if } x < 0, \\ \\ \mathbf{Q_R} & \text{if } x > 0, \end{array} \right\} \tag{44}$$

where the vector $\mathbf{Q}^{(0)}$ of initial conditions is

$$\mathbf{Q}^{(0)} = [q_1^{(0)}, \dots, q_m^{(0)}]^{\mathrm{T}}.$$

We find the general solution of the IVP (44) by first solving the corresponding IVP for the canonical system (41) or (42) in terms of the characteristic variables $\mathbf{W}$ and appropriate initial condition. We solve

$$\text{PDEs}: \partial_t \mathbf{W} + \Lambda \partial_x \mathbf{W} = \mathbf{0}, \quad -\infty < x < \infty, t > 0, \left.\begin{array}{c} \\ \\ \end{array}\right\}$$

$$\text{IC}: \quad \mathbf{W}(x,0) = \mathbf{W}^{(0)}(x) = \left\{ \begin{array}{ll} \mathbf{W_L} & \text{if } x < 0, \\ \\ \mathbf{W_R} & \text{if } x > 0, \end{array} \right\} \tag{45}$$

with

$$\mathbf{W}^{(0)} = [w_1^{(0)}, \dots, w_m^{(0)}]^{\mathrm{T}} = \mathbf{R}^{-1}\mathbf{Q}^{(0)} \text{ or } \mathbf{Q}^{(0)} = \mathbf{R}\mathbf{W}^{(0)}.$$

The solution of IVP (45) is direct. By considering each unknown $w_i(x,t)$ in (43) and its corresponding initial data $w_i^{(0)}$, we write its solution immediately as

$$w_i(x,t) = w_i^{(0)}(x - \lambda_i t), \text{ for } i = 1, \dots, m. \tag{46}$$

Compare this with the solution (15) for the scalar case. The solution of the general IVP in terms of the original variables $\mathbf{Q}$ is now obtained by transforming back according to (40), namely $\mathbf{Q} = \mathbf{R}\mathbf{W}$, to recover the solution to the original problem. When written in full the solution becomes

$$q_1 = w_1 r_1^{(1)} + w_2 r_1^{(2)} + \cdots + w_m r_1^{(m)},$$

$$q_i = w_1 r_i^{(1)} + w_2 r_i^{(2)} + \cdots + w_m r_i^{(m)},$$

$$q_m = w_1 r_m^{(1)} + w_2 r_m^{(2)} + \cdots + w_m r_m^{(m)},$$

or

$$\begin{bmatrix} q_1 \\ q_2 \\ \vdots \\ q_m \end{bmatrix} = w_1 \begin{bmatrix} r_1^{(1)} \\ r_2^{(1)} \\ \vdots \\ r_m^{(1)} \end{bmatrix} + w_2 \begin{bmatrix} r_1^{(2)} \\ r_2^{(2)} \\ \vdots \\ r_m^{(2)} \end{bmatrix} + \cdots + w_m \begin{bmatrix} r_1^{(m)} \\ r_2^{(m)} \\ \vdots \\ r_m^{(m)} \end{bmatrix},$$

or more succinctly

$$\mathbf{Q}(x,t) = \sum_{i=1}^{m} w_i(x,t)\mathbf{R}_i. \tag{47}$$

This means that the function $w_i(x,t)$ is the coefficient of $\mathbf{R}_i$ in an *eigenvector expansion* of the vector $\mathbf{Q}$. But according to (46), $w_i(x,t) = w_i^{(0)}(x - \lambda_i t)$ and hence

$$\mathbf{Q}(x,t) = \sum_{i=1}^{m} w_i^{(0)}(x - \lambda_i t)\mathbf{R}_i. \tag{48}$$

Thus, given a point (x,t) in the x–t plane, the solution $\mathbf{Q}(x,t)$ at this point depends only on the initial data at the m points $x_0^{(i)} = x - \lambda_i t$. These are the intersections of the characteristics of speed λ_i with the x-axis. The solution (48) for $\mathbf{Q}$ can be seen as the superposition of m waves, each of which is advected independently without change in shape. The ith wave has the shape $w_i^{(0)}(x)\mathbf{R}_i$ and propagates with speed λ_i.

The Riemann Problem

We study the Riemann problem for the hyperbolic, constant coefficient system (37), namely the IVP

$$\left. \begin{array}{l} \text{PDEs: } \partial_t \mathbf{Q} + \mathbf{A}\partial_x \mathbf{Q} = \mathbf{0}, \ -\infty < x < \infty, \ t > 0, \\[2mm] \text{IC: } \quad \mathbf{Q}(x,0) = \mathbf{Q}^{(0)}(x) = \begin{cases} \mathbf{Q}_{\mathrm{L}} \text{ if } x < 0, \\[2mm] \mathbf{Q}_{\mathrm{R}} \text{ if } x > 0, \end{cases} \end{array} \right\} \tag{49}$$

which is a special case of IVP (44) and a generalization of IVP (16). We assume that the system is *strictly hyperbolic* and we order the real and distinct eigenvalues as

$$\lambda_1 < \lambda_2 < \cdots < \lambda_m. \tag{50}$$

The structure of the solution of the Riemann problem (49) in the x–t plane consists of m waves emanating from the origin, one for each eigenvalue λ_i. Each wave i carries a *jump discontinuity* in $\mathbf{Q}$ propagating with speed λ_i. Naturally, the solution to the left of the λ_1-wave is simply the initial data $\mathbf{Q}_{\mathrm{L}}$ and to the right of the λ_m-wave the solution is $\mathbf{Q}_{\mathrm{R}}$. The task at hand is to find the solution in the wedge between the λ_1 and λ_m waves.

As the eigenvectors $\mathbf{R}_1, \ldots, \mathbf{R}_m$ are *linearly independent*, we can expand the data $\mathbf{Q}_{\mathrm{L}}$, constant left state, and $\mathbf{Q}_{\mathrm{R}}$, constant right state, as linear combinations of the set $\mathbf{R}_1, \ldots, \mathbf{R}_m$, that is

$$\mathbf{Q}_{\mathrm{L}} = \sum_{i=1}^{m} \alpha_i \mathbf{R}_i, \quad \mathbf{Q}_{\mathrm{R}} = \sum_{i=1}^{m} \beta_i \mathbf{R}_i, \tag{51}$$

with constant coefficients α_i and β_i, for $i = 1, \ldots, m$. Formally, the solution of the IVP (49) is given by (48) in terms of the initial data $w_i^{(0)}(x)$ for the characteristic variables and the right eigenvectors $\mathbf{R}_i$. Note that each of the expansions in (51) is a special case of (48). In terms of the characteristic variables, we have m scalar Riemann problems for the PDEs

$$
\left.
\begin{aligned}
&\text{PDE:} \quad \frac{\partial w_i}{\partial t} + \lambda_i \frac{\partial w_i}{\partial x} = 0, \quad -\infty < x < \infty, \ t > 0 \\[2mm]
&\text{IC:} \quad w_i^{(0)}(x) =
\begin{cases}
\alpha_i \ \text{if } x < 0, \\[2mm]
\beta_i \ \text{if } x > 0,
\end{cases}
\end{aligned}
\right\}
\tag{52}
$$

for $i = 1, \ldots, m$. From the previous results, see (18), we know that the solutions of these scalar Riemann problems are given by

$$
w_i(x, t) = w_i^{(0)}(x - \lambda_i t) =
\begin{cases}
\alpha_i \ \text{if } x - \lambda_i t < 0, \\[2mm]
\beta_i \ \text{if } x - \lambda_i t > 0.
\end{cases}
\tag{53}
$$

For a given point (x, t), there is an eigenvalue λ_I such that $\lambda_I < x/t < \lambda_{I+1}$, that is, $x - \lambda_i t > 0 \ \forall i$ such that $i \leq I$. We can thus write the final solution to the Riemann problem (49) in terms of the original variables as

$$
\mathbf{Q}(x, t) = \sum_{i=I+1}^{m} \alpha_i \mathbf{R}_i + \sum_{i=1}^{I} \beta_i \mathbf{R}_i,
\tag{54}
$$

where the integer $I = I(x, t)$ is the maximum value of the sub-index i for which $x - \lambda_i t > 0$.

1.4 Non-linear Equations

Consider the one-dimensional non-linear system

$$
\partial_t \mathbf{Q} + \partial_x \mathbf{F}(\mathbf{Q}) = \mathbf{S}(\mathbf{Q}),
\tag{55}
$$

where $\mathbf{Q}$ is the vector of conserved variables, $\mathbf{F}(\mathbf{Q})$ is the vector of fluxes and $\mathbf{S}(\mathbf{Q})$ is the vector of source terms.

As mentioned earlier, conservation laws may be expressed in differential or in integral form. Here we consider two variants of the integral form. Take a control volume $V = [x_L, x_R] \times [t_1, t_2]$ on the x–t plane. One integral form of the system is

$$
\frac{d}{dt} \int_{x_L}^{x_R} \mathbf{Q}(x, t) \, dx = \mathbf{F}(\mathbf{Q}(x_L, t)) - \mathbf{F}(\mathbf{Q}(x_R, t)) + \int_{x_L}^{x_R} \mathbf{S}\left(\mathbf{Q}(x, t)\right) \, dx.
\tag{56}
$$

Another version of the integral form of the conservation laws is obtained by integrating (56) in time between t_1 and t_2, with $t_1 \leq t_2$. Clearly,

$$\int_{t_1}^{t_2} \left[\frac{d}{dt} \int_{x_L}^{x_R} \mathbf{Q}(x,t)\, dx \right] dt = \int_{x_L}^{x_R} \mathbf{Q}(x,t_2)\, dx - \int_{x_L}^{x_R} \mathbf{Q}(x,t_1)\, dx$$

and thus (56) becomes

$$\left. \begin{array}{l} \displaystyle\int_{x_L}^{x_R} \mathbf{Q}(x,t_2)\, dx = \int_{x_L}^{x_R} \mathbf{Q}(x,t_1)\, dx \\[2em] \displaystyle\qquad + \int_{t_1}^{t_2} \mathbf{F}(\mathbf{Q}(x_L,t))\, dt - \int_{t_1}^{t_2} \mathbf{F}(\mathbf{Q}(x_R,t))\, dt \\[2em] \displaystyle\qquad + \int_{t_1}^{t_2} \int_{x_L}^{x_R} \mathbf{S}\left(\mathbf{Q}(x,t)\right)\, dx dt. \end{array} \right\} \qquad (57)$$

Shocks Waves and the Rankine–Hugoniot Conditions

Given a system of hyperbolic conservation laws

$$\partial_t \mathbf{Q} + \partial_x \mathbf{F}(\mathbf{Q}) = \mathbf{0} \qquad (58)$$

and a discontinuous solution of speed S_i associated with the λ_i-characteristic field, the Rankine–Hugoniot conditions state

$$\Delta \mathbf{F} = S_i \Delta \mathbf{Q}, \qquad (59)$$

with

$$\Delta \mathbf{Q} \equiv \mathbf{Q}_R - \mathbf{Q}_L, \;\; \Delta \mathbf{F} \equiv \mathbf{F}_R - \mathbf{F}_L, \;\; \mathbf{F}_L = \mathbf{F}(\mathbf{Q}_L), \;\; \mathbf{F}_R = \mathbf{F}(\mathbf{Q}_R),$$

where $\mathbf{Q}_L$ and $\mathbf{Q}_R$ are the respective states immediately to the left and right of the discontinuity.

Example. Burgers's equation. Assume a shock wave of speed s with states q_L and q_R modelled by the inviscid Burgers equation (6). The Rankine–Hugoniot condition give

$$f_R - f_L = s(q_R - q_L), \qquad (60)$$

$$\frac{1}{2}q_R^2 - \frac{1}{2}q_L^2 = s(q_R - q_L), \qquad (61)$$

from which the shock speed is given by

$$s = \frac{1}{2}(q_L + q_R). \tag{62}$$

Note that, unlike the scalar case, it is generally not possible to solve for the shock speed S_i in the case of a system. For a linear system with constant coefficients

$$\partial_t \mathbf{Q} + \mathbf{A}\partial_x \mathbf{Q} = \mathbf{0},$$

with eigenvalues λ_i, for $i = 1, \ldots, m$, the Rankine–Hugoniot conditions across the wave of speed $S_i \equiv \lambda_i$ read

$$\Delta \mathbf{F} = \mathbf{A}\Delta \mathbf{Q} = \lambda_i (\Delta \mathbf{Q})_i. \tag{63}$$

In this brief introduction to hyperbolic balance laws, we have omitted the study of non-linear equations and issues such as shock formation, non-uniqueness, entropy conditions and many other concepts. See the guidelines for further reading.

1.5 Further Reading

An informal introduction to the theory of hyperbolic conservation laws is found in Chap. 2 of [36]. Also, an introductory treatment of hyperbolic equations is found in Chap. 10 of [46]. Treatments of the theory of conservation laws with numerical methods in mind are found in [21] and [13]. For the theoretically inclined reader, comprehensive treatments of the theory of hyperbolic of conservation laws are found in [5, 27] and [20].

2 The Finite Difference Method

A succinct introduction to numerical methods for computing approximate solutions to hyperbolic equations is presented. Most of the section is devoted to finite difference methods and some of their basic properties. We also introduce finite volume methods, as derived from the integral form of the equations on control volumes, and point out some of their properties. A list of relevant references for further study is also given at the end of the section.

We study some basic concepts on numerical methods for hyperbolic partial differential equations (PDEs). We do so in terms of the initial-boundary value problem for the model hyperbolic equation, namely

$$\left.\begin{array}{l} \text{PDE:}\ \ \partial_t q + \lambda \partial_x q = 0, \ \ x \in (0, b),\ t > 0, \\[2mm] \text{IC:}\ \ q(x, 0) = q^{(0)}(x), \ \ x \in (0, b), \\[2mm] \text{BCs:}\ \ q(0, t) = q_0(t), \ \ q(b, t) = q_b(t),\ t \geq 0, \end{array}\right\} \tag{64}$$

where b is a positive real number; $q^{(0)}(x)$ is the initial condition (IC) of the problem, a prescribed function of x; and $q_0(t)$ and $q_b(t)$ are prescribed functions of time and define the boundary conditions (BCs) of the problem.

The finite difference method first replaces the continuous x–t domain of problem (64) by a discrete domain, a finite set of points (x_i, t_n) called *mesh* or *grid*. Then, at each point (x_i, t_n) the partial derivatives of the PDE are substituted by finite difference approximations; in this manner the partial differential equation is substituted by a difference equation, an expression that relates approximate values of the solution at neighbouring points.

Generating a mesh, specially for complex domains in multiple space dimensions, can be a very demanding task. In its simplest form, as for problem (64), the mesh can be generated as follows:

- the spatial domain $[0, b]$ is partitioned by a set of $M + 2$ equidistant points

$$x_i = i\Delta x, i = 0, \ldots, M, \Delta x = \frac{b}{M + 1}, \tag{65}$$

where M is a chosen positive integer and we end up with M interior points and two boundary points.
- the temporal domain $[0, \infty)$ is partitioned by a set of equidistant points, or time levels,

$$t_n = n\Delta t, n = 0, \ldots, \tag{66}$$

We end up with the set of points $(x_i, t_n) = (i\Delta x, n\Delta t)$, also denoted by (i, n). The parameters Δx and Δt determine the mesh. Δx results from prescribing the number of points in the partition of $[0, b]$, while Δt is related to Δx in a fixed manner, as we shall see later. Δx is the distance between points in the x-direction and Δt is the *time step*, the difference between two time levels.

2.1 Finite Difference Approximation to Derivatives

The finite difference method computes an approximate value q_i^n to the exact value $q(x_i, t_n)$ of the solution $q(x, t)$ of (64) at a finite set of points (x_i, t_n) in the x–t domain of problem (64). Thus we write

$$q_i^n \approx q(x_i, t_n). \tag{67}$$

For example, at the point (x_i, t_n), the temporal partial derivative can be replaced by any of the following expressions:

$$\partial_t q(x_i, t_n) = \left\{ \begin{array}{l} \dfrac{q_i^{n+1} - q_i^n}{\Delta t} + \mathcal{O}(\Delta t), \\[3mm] \dfrac{q_i^{n+1} - q_i^{n-1}}{2\Delta t} + \mathcal{O}(\Delta t^2). \end{array} \right\} \tag{68}$$

Analogously for the spatial partial derivative in (64) at the point (x_i, t_n), we can use any of the following:

$$\partial_x q(x_i, t_n) = \begin{cases} \frac{q_i^n - q_{i-1}^n}{\Delta x} + \mathcal{O}(\Delta x), \\[2mm] \frac{q_{i+1}^n - q_i^n}{\Delta x} + \mathcal{O}(\Delta x), \\[2mm] \frac{q_{i+1}^n - q_{i-1}^n}{2\Delta x} + \mathcal{O}(\Delta x^2). \end{cases} \tag{69}$$

Other choices for the finite difference approximations are of course possible.

2.2 Some Well-Known Numerical Methods

Particular numerical methods result from choosing particular combinations in (68) and (69) above, not all of them being productive.

Godunov's Method (1959). This method, interpreted in a finite difference setting, results from choosing

$$\partial_t q(x_i, t_n) = \frac{q_i^{n+1} - q_i^n}{\Delta t},$$

$$\partial_x q(x_i, t_n) = \begin{cases} \frac{q_i^n - q_{i-1}^n}{\Delta x} & \text{if } \lambda > 0, \\[2mm] \frac{q_{i+1}^n - q_i^n}{\Delta x} & \text{if } \lambda < 0. \end{cases} \tag{70}$$

For positive characteristic speed, $\lambda > 0$, after substituting the finite difference approximations (70) into the PDE in (64), we obtain

$$L_a(q_i^n) \equiv \frac{q_i^{n+1} - q_i^n}{\Delta t} + \lambda \left(\frac{q_i^n - q_{i-1}^n}{\Delta x} \right) = 0, \tag{71}$$

where $L_a(q_i^n)$ denotes a numerical operator associated with the Godunov method and acts on point values. Solving for q_i^{n+1}, we obtain the numerical scheme

$$q_i^{n+1} = q_i^n - c\left(q_i^n - q_{i-1}^n \right), \tag{72}$$

where

$$c = \frac{\lambda \Delta t}{\Delta x} = \frac{\lambda}{\Delta x / \Delta t} \tag{73}$$

is the Courant number, or the CFL number (for Courant–Friedrichs–Lewy). Note that c is a dimensionless quantity, as is the ratio of the characteristic speed λ in the PDE and the *mesh speed* $\Delta x / \Delta t$.

For negative characteristic speed, $\lambda < 0$, the Godunov scheme reads

$$q_i^{n+1} = q_i^n - c(q_{i+1}^n - q_i^n). \tag{74}$$

Remarks

- The method (72), or (74), is said to be *explicit* as the solution at the new time level $n + 1$ at the point i depends explicitly on the solution at the previous time level n.
- Formulas (72) and (74) are *time-marching* procedures. Solution values at the future time level $n + 1$ can be *predicted* by using the known solution at the time level n (the present time).
- At time level 0, one uses the initial condition $q^{(0)}(x)$ to provide the discrete initial values $q_i^0 = q^{(0)}(x_i)$.
- The mesh points that lie on the boundaries $x = 0$ and $x = b$ must be updated separately, using the boundary conditions of the problem. Beware of boundary conditions.
- The scheme (72), and (74), is said to be a *one-step scheme*. It is also called a *two-level scheme*.
- The Godunov method is said to be *upwind*, as the spatial differencing in (70) is performed according to the sign of the characteristic speed λ. The information is taken from the side from which the *wind blows*.

The FTCS (Forward in Time Central in Space) Method. This finite difference method results from choosing

$$\left. \begin{array}{l} \partial_t q(x_i, t_n) = \frac{q_i^{n+1} - q_i^n}{\Delta t}, \\[2mm] \partial_x q(x_i, t_n) = \frac{q_{i+1}^n - q_{i-1}^n}{2\Delta x}. \end{array} \right\} \tag{75}$$

The approximate operator is

$$L_a(q_i^n) \equiv \frac{q_i^{n+1} - q_i^n}{\Delta t} + \lambda \left(\frac{q_{i+1}^n - q_{i-1}^n}{2\Delta x} \right) = 0, \tag{76}$$

which gives the explicit scheme

$$q_i^{n+1} = q_i^n - \frac{1}{2} c(q_{i+1}^n - q_{i-1}^n). \tag{77}$$

As we shall see later, the FTCS method is unconditionally unstable and is thus useless.

The Lax–Friedrichs Method. This finite difference method results from choosing finite difference approximations as in the FTCS method but replacing q_i^n in the approximation to the time derivative by the mean value $1/2(q_{i-1}^n + q_{i+1}^n)$. The numerical operator is

$$L_a(q_i^n) \equiv \frac{q_i^{n+1} - \frac{1}{2}(q_{i-1}^n + q_{i+1}^n)}{\Delta t} + \lambda \left(\frac{q_{i+1}^n - q_{i-1}^n}{2\Delta x} \right) = 0, \tag{78}$$

yielding the explicit scheme

$$q_i^{n+1} = \frac{1}{2}(q_{i-1}^n + q_{i+1}^n) - \frac{1}{2}c(q_{i+1}^n - q_{i-1}^n). \tag{79}$$

The Lax–Wendroff Method. The construction of this method is somewhat different. One first expresses the solution at (x_i, t_{n+1}) as a Taylor series expansion in time

$$q(x_i, t_{n+1}) = q(x_i, t_n) + \Delta t \partial_t q(x_i, t_n) + \frac{1}{2}\Delta t^2 \partial_t^{(2)} q(x_i, t_n) + \mathcal{O}(\Delta t^3). \tag{80}$$

By means of the Cauchy–Kowalewski procedure, one can use the PDE in (64) to replace time derivatives by space derivatives, namely

$$\left.\begin{aligned}
\partial_t q(x,t) &= -\lambda \partial_x q(x,t), \\
\partial_t^{(2)} q(x,t) &= \lambda^2 \partial_x^{(2)} q(x,t), \\
\partial_t^{(k)} q(x,t) &= (-\lambda)^k \partial_x^{(k)} q(x,t).
\end{aligned}\right\} \tag{81}$$

Then we have

$$q(x_i, t_{n+1}) = q(x_i, t_n) - \Delta t \lambda \partial_x q(x_i, t_n) + \frac{1}{2}\Delta t^2 \lambda^2 \partial_x^{(2)} q(x_i, t_n) + \mathcal{O}(\Delta t^3). \tag{82}$$

Now by approximating the spatial derivatives by central finite differences

$$\partial_x q(x_i, t_n) = \frac{q_{i+1}^n - q_{i-1}^n}{2\Delta x}, \quad \partial_x^{(2)} q(x_i, t_n) = \frac{q_{i+1}^n - 2q_i^n + q_{i-1}^n}{\Delta x^2} \tag{83}$$

and substituting exact values by their respective approximate values we obtain the Lax–Wendroff scheme

$$q_i^{n+1} = \frac{1}{2}c(1+c)q_{i-1}^n + (1-c^2)q_i^n - \frac{1}{2}c(1-c)q_{i+1}^n. \tag{84}$$

2.3 Some Properties of Numerical Methods

It is intuitively obvious that one would expect the numerical approximation to the PDE to converge to the exact solution of the PDE as the mesh is refined, that is, as the mesh parameters Δx and Δt tend to zero. Mathematically, however, it is not a simple matter to demonstrate directly that a particular method is convergent. However, by means of the *Lax Equivalent Theorem*, one can indirectly arrive at the result. The *Lax Equivalent Theorem*, valid for linear problems, states that a method is convergent if and only if it is *consistent* and *stable*. We therefore study here, these two latter concepts.

All schemes studied so far can be written as

$$q_i^{n+1} = H(q_{i-l}^n, q_{i-l+1}^n, \ldots, q_i^n, \ldots, q_{i+r}^n), \tag{85}$$

where l and r are two non-negative integers that determine the support of the scheme. For example, for the Godunov method, the operator H is in fact a linear combination of solution values at the time level n, namely

$$
\left.\begin{array}{l}
H = cq_{i-1}^n + (1-c)q_i^n, \text{ for } \lambda > 0 , \ \ l = -1, \ \ r = 0, \\[2mm]
H = (1+c)q_i^n - cq_{i+1}^n, \text{ for } \lambda < 0 , \ \ l = 0, \ \ r = 1.
\end{array}\right\} \tag{86}
$$

Linear Schemes. A numerical scheme, for the linear advection equation with constant coefficient λ in (64), written as a linear combination of data values at time level n

$$
q_i^{n+1} = \sum_{k=-l}^{k=r} b_k q_{i+k}^n \tag{87}
$$

is said to be *linear* if the coefficients b_k are constant, that is, they do not depend on the solution. All schemes considered so far are examples on linear schemes.

Local Truncation Error

Amongst the several types of errors of interest, the local truncation error of a numerical scheme is of basic importance. It measures, locally, the quality of the approximation of the PDE by a difference equation in one time step. One begins by expressing the scheme (85) as the approximate operator $L_a(q_i^n)$ introduced in (71), namely

$$
L_a(q_i^n) = \frac{1}{\Delta t}[q_i^{n+1} - H(q_{i-l}^n, q_{i-l+1}^n, \ldots, q_i^n, \ldots, q_{i+r}^n)] = 0. \tag{88}
$$

In this form, the numerical operator $L_a(q_i^n)$ is the discrete analogue of the differential operator. Note that $L_a(q_i^n) = 0$, that is, q_i^n satisfies identically the difference equation. But how about $L_a(q(x_i, t_n))$? Is the exact solution $q(x_i, t_n)$ of the PDE at the point (x_i, t_n) also a solution of the discrete problem? The answer is no, in general. There will be an error and this is called the local truncation error, defined as

$$
T_i^n = L_a(q(x_i, t_n)). \tag{89}
$$

The fact that T_i^n is not zero is not due to the argument $q(x_i, t_n)$ in (89), as this is the exact solution of the PDE. It is due to the approximate operator, and thus the local truncation error is a measure of the numerical scheme.

The general procedure to calculate the truncation error first assumes that the solution $q(x, t)$ of the PDE is sufficiently smooth so that one can use Taylor series expansions, in space and time, about the central point of the stencil (x_i, t_n). Then, algebraic manipulations lead to an expression of the form

$$T_i^n = \partial_t q(i\Delta x, n\Delta t) + \lambda \partial_x q(i\Delta x, n\Delta t) \Bigg\}$$
$$+ \; \mathcal{O}(\Delta t^k) + \mathcal{O}(\Delta x^m). \tag{90}$$

The first line of the right hand side of (90) is precisely the differential equation of which $q(x, t)$ is the solution. Therefore this line is zero and the local truncation error has the form

$$T_i^n = \mathcal{O}(\Delta t^k) + \mathcal{O}(\Delta x^m). \tag{91}$$

Order of a Method. We say that the scheme (85) with local truncation error (91) is of the order k in time and of the order m in space. The order of accuracy p of scheme (85) is defined as $p = min\{k, m\}$.

Example. **Local Truncation Error of Godunov's Method.** Applying the definition of local truncation error, we have

$$T_i^n = \frac{1}{\Delta t} \left[q(i\Delta x, (n+1)\Delta t) - (cq((i-1)\Delta x, n\Delta t) + (1-c)q(i\Delta x, n\Delta t)) \right]. \tag{92}$$

Assuming the solution $q(x, t)$ to be smooth, we develop the following Taylor expansions about the point (x_i, t_n).

$$
\left.
\begin{aligned}
q(i\Delta x, (n+1)\Delta t) &= q(i\Delta x, n\Delta t) + \Delta t \partial_t q(i\Delta x, n\Delta t) \\
&\quad + \tfrac{1}{2}\Delta t^2 \partial_t^{(2)} q(i\Delta x, n\Delta t) + \mathcal{O}(\Delta t^3), \\[4pt]
q((i-1)\Delta x, n\Delta t) &= q(i\Delta x, n\Delta t) - \Delta x \partial_x q(i\Delta x, n\Delta t) \\
&\quad + \tfrac{1}{2}\Delta x^2 \partial_x^{(2)} q(i\Delta x, n\Delta t) + \mathcal{O}(\Delta x^3), \\[4pt]
q((i+1)\Delta x, n\Delta t) &= q(i\Delta x, n\Delta t) + \Delta x \partial_x q(i\Delta x, n\Delta t) \\
&\quad + \tfrac{1}{2}\Delta x^2 \partial_x^{(2)} q(i\Delta x, n\Delta t) + \mathcal{O}(\Delta x^3).
\end{aligned}
\right\} \tag{93}
$$

Substituting (93) into (92) and using the fact that $c\Delta x/\Delta t = \lambda$ and $c\Delta x^2/\Delta t = \lambda \Delta x$, we obtain

$$
\left.
\begin{aligned}
T_i^n &= \partial_t q(i\Delta x, n\Delta t) + \lambda \partial_x q(i\Delta x, n\Delta t) \\[4pt]
&\quad + \tfrac{1}{2}\Delta t \partial_t^{(2)} q(i\Delta x, n\Delta t) - \tfrac{1}{2}\lambda \Delta x \partial_x^{(2)} q(i\Delta x, n\Delta t) + \mathcal{O}(\Delta t^2) + \mathcal{O}(\Delta x^2).
\end{aligned}
\right\} \tag{94}
$$

Note that the first line of the right-hand side of (94) is precisely the differential equation, of which $q(x, t)$ it is assumed to be the solution. Consequently, this line is zero. Then neglecting the terms $\mathcal{O}(\Delta t^2)$ and $\mathcal{O}(\Delta x^2)$, we obtain the local truncation error

$$T_i^n = \frac{1}{2}\Delta t \partial_t^{(2)} q(i\Delta x, n\Delta t) - \frac{1}{2}\lambda \Delta x \partial_x^{(2)} q(i\Delta x, n\Delta t) = \mathcal{O}(\Delta t) + \mathcal{O}(\Delta x). \tag{95}$$

As $T_i^n = \mathcal{O}(\Delta t) + \mathcal{O}(\Delta x)$ the Godunov scheme is said to be first-order in space and first-order in time.

Now we use the Cauchy–Kowalewski procedure in (95) to convert the time derivative to a space derivative, see (81), and we obtain

$$T_i^n = \frac{1}{2}\lambda\Delta x(c-1)\partial_x^{(2)}q(i\Delta x, n\Delta t). \tag{96}$$

Remark. The local truncation error for the Godunov method vanishes for $\Delta x = 0$, as one would expect, for $\lambda = 0$ and also for $c = 1$, Courant number unity. Note also that the error depends on a second-order spatial derivative.

Consistent Scheme. A numerical scheme is said to be consistent with the PDE (or compatible with the PDE) if the truncation error vanishes as the mesh parameters Δx and Δt tend to zero.

Modified Equation

In general, a numerical method does not solve the PDE that one intends to solve, but due to truncation errors inherent in the scheme, one solves more accurately other PDEs. Such PDEs may be seen as variations of the original PDE in which the local truncation error plays a major role.

As a first example, consider the local truncation error of the Godunov upwind method, for $\lambda > 0$,

$$\left.\begin{aligned}T_i^n = {}& \partial_t q(i\Delta x, n\Delta t) + \lambda\partial_x q(i\Delta x, n\Delta t) \\ &+ \tfrac{1}{2}[\Delta t\partial_t^{(2)}q(i\Delta x, n\Delta t) - \lambda\Delta x\partial_x^{(2)}q(i\Delta x, n\Delta t)] + \mathcal{O}(\Delta t^2) + \mathcal{O}(\Delta x^2).\end{aligned}\right\} \tag{97}$$

In calculating the local truncation error, we have assumed that $q(x,t)$ is a solution of the original PDE and we have thus neglected the first line on the right hand side of (97). Had we assumed that $q(x,t)$ was a solution of the *modified* PDE at the point $(i\Delta x, n\Delta t)$

$$\partial_t q + \lambda\partial_x q + \frac{1}{2}[\Delta t\partial_t^{(2)}q - \lambda\Delta x\partial_x^{(2)}q] = 0 \tag{98}$$

then the local truncation error would have been of the order $\mathcal{O}(\Delta t^2), \mathcal{O}(\Delta x^2)$. In this case, the scheme solves the modified equation (98) more accurately than the original PDE, to second order, in fact.

The modified equation for the Godunov upwind scheme can be written as

$$\partial_t q + \lambda\partial_x q = \frac{1}{2}\left[\lambda\Delta x\partial_x^{(2)}q - \Delta t\partial_t^{(2)}q\right]. \tag{99}$$

Assuming a fixed relation between the mesh parameters Δx and Δt, we use $\mathcal{O}(\Delta t^m)$ also for $\mathcal{O}(\Delta x^m)$. Now applying the Cauchy–Kowalewski procedure to the modified equation (99) to convert the second-order time derivative to a space derivative, we have

$$\left.\begin{aligned}
\partial_t q &= -\lambda \partial_x q = \tfrac{1}{2}[\lambda \Delta x \partial_x^{(2)} q - \Delta t \partial_t^{(2)} q] = -\lambda \partial_x q + \mathcal{O}(\Delta t), \\
\partial_x \partial_t q &= -\lambda \partial_x^2 q + \mathcal{O}(\Delta t), \\
\partial_t^2 q &= -\lambda(\partial_x \partial_t q + \mathcal{O}(\Delta t)) = -\lambda[-\lambda \partial_x^2 q + \mathcal{O}(\Delta t)] + \mathcal{O}(\Delta t), \\
\partial_t^2 q &= \lambda^2 \partial_x^2 q + \mathcal{O}(\Delta t).
\end{aligned}\right\} \tag{100}$$

The modified equation becomes

$$\left.\begin{aligned}
\partial_t q + \lambda \partial_x q &= \alpha \partial_x^{(2)} q, \\
\alpha &= \tfrac{1}{2}\lambda \Delta x (1 - c).
\end{aligned}\right\} \tag{101}$$

This is an advection–diffusion equation, a parabolic equation. Numerical solutions of the inviscid equation behave as the solution of a viscous equation, having the effect of *numerical viscosity*, not of *physical viscosity*. For this reason, the coefficient α is called the coefficient of *numerical viscosity*. In general, first-order methods have modified equations that are of the advection–diffusion type, with *numerical viscosity*.

Second-order methods, on the other hand, have modified equations of the dispersion type, namely

$$\partial_t q + \lambda \partial_x q = \gamma \partial_x^{(3)} q, \tag{102}$$

with coefficient of *numerical dispersion* γ. For example, the Lax–Wendroff method, a second-order method, has modified equation

$$\left.\begin{aligned}
\partial_t q + \lambda \partial_x q &= \alpha \partial_x^{(3)} q, \\
\gamma &= \tfrac{1}{6}\lambda \Delta x^2 (c^2 - 1).
\end{aligned}\right\} \tag{103}$$

Numerical dispersive errors show up in the form of wrong wave propagation speeds, called phase errors.

Linear Stability Analysis

Here we deal with stability analysis of numerical schemes. There are several views on the meaning of stability, or instability, but a central issue is that of *spurious oscillations and their unbounded growth in time*. An informative discussion is found in the book by Laney [19]. A distinction must be made between linear and non-linear stability. Here we are concerned exclusively with linear stability, for which again there are several methods to analyse it. We study the von Neumann method and apply it, as an example, to the Godunov scheme (72). We introduce the Fourier component, or *trial solution*,

$$q_i^n = A^n e^{I\theta i}, \tag{104}$$

where A is the amplitude (raised to the power n), a real or complex number, $\theta = P\Delta x$ is an angle, P is the wave number in the x-direction and we define $I = \sqrt{-1}$, to avoid confusion with the spatial index i. Substitution of the trial solution into the numerical scheme

$$q_i^{n+1} = q_i^n - c\left(q_i^n - q_{i-1}^n\right) \tag{105}$$

gives

$$A^{n+1} e^{I\theta i} = A^n e^{I\theta i} - c\left(A^n e^{I\theta i} - A^n e^{I\theta(i-1)}\right). \tag{106}$$

Cancelling the common factor $A^n e^{I\theta i}$ yields

$$A = 1 + c(\cos\theta - 1) - cI\sin\theta. \tag{107}$$

The squared of the modulus of this complex number is

$$||A||^2 = \left(1 + c(\cos\theta - 1)\right)^2 + c^2\sin^2\theta. \tag{108}$$

Manipulations give

$$||A||^2 = 1 - 2c(1-c)(1 - \cos\theta), \tag{109}$$

from which it follows that

$$||A||^2 \leq 1 \text{ if } 0 \leq c \leq 1. \tag{110}$$

The Godunov scheme is stable if condition (110) is satisfied. We say that the Godunov scheme is *conditionally stable* with stability condition (110).

Monotonicity and Accuracy

Monotone scheme. A numerical scheme

$$q_i^{n+1} = H(q_{i-l}^n, q_{i-l+1}^n, \ldots, q_i^n, \ldots, q_{i+r}^n) \tag{111}$$

is said to be monotone if

$$\frac{\partial}{\partial q_k^n} H(q_{i-l}^n, q_{i-l+1}^n, \ldots, q_i^n, \ldots, q_{i+r}^n) \geq 0, \quad i - l \leq k \leq i + r. \tag{112}$$

Note that for a linear scheme

$$q_i^{n+1} = \sum_{k=-l}^{k=r} b_k q_{i+k}^n,\tag{113}$$

where the coefficients b_k are constant, monotonicity requires that all coefficients be non-negative, that is

$$b_k \geq 0, \quad i - l \leq k \leq i + r.\tag{114}$$

Thus, by inspection one can decide whether a numerical scheme is a monotone or not. For example, the Godunov upwind method for $\lambda > 0$ is

$$q_i^{n+1} = H(q_{i-l}^n, q_i^n) = c q_{i-l}^n + (1 - c) q_i^n.\tag{115}$$

Clearly the scheme is monotone, as both coefficients are non-negative if the linear stability condition (110) is obeyed, that is, $c \geq 0$ and $1 - c \geq 0$, if $0 \leq c \leq 1$.

Exercise. Discuss the monotonicity of the following schemes: Godunov's scheme for $\lambda < 0$, the FTCS scheme, the Lax–Friedrichs scheme and the Lax–Wendroff scheme.

We now state a useful result that facilitates the verification of the accuracy of any linear scheme

$$q_i^{n+1} = \sum_{k=-l}^{k=r} b_k q_{i+k}^n.\tag{116}$$

Accuracy Theorem: A scheme of the form (116) is pth ($p \geq 0$) order accurate in space and time if and only if

$$\sum_{k=-l}^{r} k^\eta b_k = (-c)^\eta, \quad 0 \leq \eta \leq p.\tag{117}$$

Remark. For each integer value of η, with $0 \leq \eta \leq p$, we verify that the sum of terms $k^\eta b_k$, for all integers k with $-l \leq k \leq r$, reproduces identically the power $(-c)^\eta$, where c is the Courant number.

The above result is used to prove the Godunov's theorem below.

Godunov's Theorem

Godunov's theorem establishes, theoretically, that the desirable properties of accuracy and monotonicity are, for linear schemes, contradictory requirements. The following result applies to the linear advection equation with constant coefficient and linear schemes (116).

Theorem. *(Godunov, 1959): There are no monotone, linear schemes (116) for the linear advection equation with constant coefficient λ of second or higher order of accuracy.*

Proof. The proof given here is based on the accuracy relation (117). Denote by s_η the summation

$$s_\eta = \sum_{k=-l}^{r} k^\eta b_k, \tag{118}$$

where b_k are the coefficients (constant) of the (linear) scheme (116). For second-order accuracy, one requires

$$s_0 = 1, \quad s_1 = -c, \quad s_2 = c^2. \tag{119}$$

From definition (118)

$$\left.\begin{aligned}
s_2 &= \sum_{k=-l}^{r} k^2 b_k \\
&= \sum_{k=-l}^{r} (k+c)^2 b_k - 2c \sum_{k=-l}^{r} k b_k - c^2 \sum_{k=-l}^{r} b_k \\
&= \left[\sum_{k=-l}^{r} (k+c)^2 b_k \right] - 2cs_1 - c^2 s_0.
\end{aligned}\right\} \tag{120}$$

Use of (119) into (120) gives

$$\left[\sum_{k=-l}^{r} (k+c)^2 b_k \right] + c^2 \geq c^2. \tag{121}$$

The above inequality holds, as a linear monotone scheme satisfies $b_k \geq 0$. Equality in (121), and thus second-order accuracy, is only possible if $b_k = 0$ $\forall k$ or when $c = -k_0$, that is for integer Courant numbers, and $b_k = 0$ $\forall k \neq k_0$. Thus the theorem has been proved for schemes satisfying the condition $0 \leq |c| \leq 1$; we note that the case of integer Courant numbers larger than unity is only of theoretical interest, as for non-linear systems this is an impossible requirement to impose on numerical methods.

Remarks on Godunov's Theorem. Another way to express Godunov's theorem is that monotone schemes are at most first-order accurate. First-order methods are too inaccurate to be of practical interest. One must therefore search for other classes of schemes that, ideally, allow for both the oscillation-free property of monotone schemes and the accuracy of high-order methods to coexist. This is down to finding ways of circumventing Godunov's theorem.

The key to this lies on the assumption made in the theorem that the schemes have fixed coefficients (linear schemes). Thus a necessary condition (not sufficient) for a numerical schemes to be oscillation-free and of high-order of accuracy (for smooth solutions) is that the scheme be *non-linear, even when applied to linear problems*. For a fuller discussion, see Chap. 13 of [36].

2.4 The Finite Volume Method

The finite volume method offers another approach to solving hyperbolic partial differential equations. Unlike the finite difference method, in which one seeks approximate values of the solution at points, the finite volume method seeks approximations to integral averages of the solution on control volumes.

The Framework

We consider the initial-boundary value problem for non-linear hyperbolic systems in one space dimension with source terms

$$
\left.
\begin{aligned}
\text{PDE: } & \partial_t \mathbf{Q} + \partial_x \mathbf{F}(\mathbf{Q}) = \mathbf{S}(\mathbf{Q}), x \in (0, b), \ t > 0, \\
\text{IC: } & \mathbf{Q}(x,0) = \mathbf{Q}^{(0)}(x), x \in (0, b), \\
\text{BCs: } & \mathbf{Q}(0,t) = \mathbf{Q}_0(t), \mathbf{Q}(b,t) = \mathbf{Q}_b(t), \ t \geq 0,
\end{aligned}
\right\} \tag{122}
$$

The finite volume method replaces the continuous spatial domain in (122) by a discrete domain consisting of a finite number of *volumes* or *cells*. Then, in each such cell one looks for approximations to the integral average of the solution. The simplest finite volume discretization of the complete x–t domain proceeds as follows:

- the spatial domain $[0, b]$ is partitioned by a set of M cells or volumes

$$
I_i \equiv [x_{i-\frac{1}{2}}, x_{i+\frac{1}{2}}] \quad , \ i = 1, \ldots, M. \tag{123}
$$

- the open temporal domain $[0, \infty)$ is partitioned by a set of time levels

$$
t_0 = 0, t_{n+1} = t_n + \Delta t_n, n = 0, \ldots \tag{124}
$$

In this setting we have the following:

$$
\left.
\begin{aligned}
& x_{i-\frac{1}{2}}, \ x_{i+\frac{1}{2}} \quad \text{are the cell interfaces,} \\
& \Delta x = x_{i+\frac{1}{2}} - x_{i-\frac{1}{2}} \quad \text{is the mesh width or cell width,} \\
& x_i = \tfrac{1}{2}(x_{i-\frac{1}{2}} + x_{i+\frac{1}{2}}) \quad \text{is the cell centre,} \\
& \Delta t_n = t_{n+1} - t_n \quad \text{is the time step or step length.}
\end{aligned}
\right\} \tag{125}
$$

Note that in practice the time step Δt_n varies from time level to time level, but for convenience we often drop the sub-index.

Consider now a space–time control volume

$$V \equiv [x_{i-\frac{1}{2}}, x_{i+\frac{1}{2}}] \times [t_n, t_{n+1}]. \tag{126}$$

Integration of the PDEs in (122) in V with respect to x and t yields the *exact* formula

$$\mathbf{Q}_i^{n+1} = \mathbf{Q}_i^n - \frac{\Delta t}{\Delta x}[\mathbf{F}_{i+\frac{1}{2}} - \mathbf{F}_{i-\frac{1}{2}}] + \Delta t \mathbf{S}_i, \tag{127}$$

with the following definitions

- $\mathbf{Q}_i^n$ is the spatial-integral average at time $t = t_n$

$$\mathbf{Q}_i^n = \frac{1}{\Delta x} \int\limits_{x_{i-\frac{1}{2}}}^{x_{i+\frac{1}{2}}} \mathbf{Q}(x, t_n)dx, \tag{128}$$

- $\mathbf{F}_{i+\frac{1}{2}}$ is the time-integral average at the interface $x = x_{i+\frac{1}{2}}$

$$\mathbf{F}_{i+\frac{1}{2}} = \frac{1}{\Delta t} \int\limits_{0}^{\Delta t} \mathbf{F}(\mathbf{Q}(x_{i+\frac{1}{2}}, t))dt, \tag{129}$$

- $\mathbf{S}_i$ is the volume-integral average in V

$$\mathbf{S}_i = \frac{1}{\Delta t} \frac{1}{\Delta x} \int\limits_{0}^{\Delta t} \int\limits_{x_{i-\frac{1}{2}}}^{x_{i+\frac{1}{2}}} \mathbf{S}(\mathbf{Q}_i(x, t))dx dt. \tag{130}$$

Remark: local coordinates. We often use local coordinates such that t_n corresponds to $t = 0$ and t_{n+1} corresponds to $t = \Delta t_n = \Delta t$. Analogously for the spatial coordinate, we take the interface position $x = x_{i+\frac{1}{2}}$ to coincide with $x = 0$.

In (128) the integrand is assumed to be the known initial condition at time t_n within the control volume. In (129) $\mathbf{Q}(x_{i+\frac{1}{2}}, t)$ is the solution of the PDEs in (122) at the interface position $x = x_{i+\frac{1}{2}}$. In (130) the function $\mathbf{Q}_i(x, t)$ is the solution of the PDEs in (122) within the space–time volume V. Under these assumptions relation (127) is exact; it is not a difference approximation.

The finite volume numerical method results from interpreting (127) as a numerical formula to update approximations to cell averages $\mathbf{Q}_i^n$. One requires approximations to the time-integral average of the flux $\mathbf{F}_{i+\frac{1}{2}}$ and approximations to the volume integral average of the source $\mathbf{S}_i$.

We call such approximations the *numerical flux* and the *numerical source*, respectively, and are still denoted by $\mathbf{F}_{i+\frac{1}{2}}$ and $\mathbf{S}_i$. There are many ways of prescribing numerical fluxes and numerical sources, and thus of obtaining finite volume methods, a large class of numerical methods.

Conservative Schemes

A conservative numerical method to solve the homogeneous version of (122) is defined as a scheme of the form

$$\mathbf{Q}_i^{n+1} = \mathbf{Q}_i^n - \frac{\Delta t}{\Delta x}[\mathbf{F}_{i+\frac{1}{2}} - \mathbf{F}_{i-\frac{1}{2}}], \tag{131}$$

in which the numerical flux is given as

$$\mathbf{F}_{i+\frac{1}{2}} = \mathbf{F}_{i+\frac{1}{2}}(\mathbf{Q}_{i-l}^n, \ldots, \mathbf{Q}_i^n, \mathbf{Q}_{i+1}^n, \ldots, \mathbf{Q}_{i+r}^n), \tag{132}$$

where l and r are two non-negative integers. As defined, the scheme is explicit. In an implicit conservative scheme, the flux function includes arguments at time level $n+1$. Finite volume schemes are, by definition, conservative methods. Some conventional finite difference methods can also be reinterpreted as conservative methods. For example, the Lax–Friedrichs method can be written as (131) with numerical flux

$$\mathbf{F}_{i+\frac{1}{2}}^{\mathrm{LF}}(\Delta x, \Delta t, \mathbf{Q}_i^n, \mathbf{Q}_{i+1}^n) = \frac{1}{2}\left(\mathbf{F}(\mathbf{Q}_i^n) + \mathbf{F}(\mathbf{Q}_{i+1}^n)\right) - \frac{1}{2}\frac{\Delta x}{\Delta t}\left(\mathbf{Q}_{i+1}^n - \mathbf{Q}_i^n\right). \tag{133}$$

The two-step version of the Lax–Wendroff method reads

$$\left.\begin{array}{l} \mathbf{F}_{i+\frac{1}{2}}^{\mathrm{LW}}(\Delta x, \Delta t, \mathbf{Q}_i^n, \mathbf{Q}_{i+1}^n) = \mathbf{F}(\mathbf{Q}_{i+\frac{1}{2}}^{\mathrm{LW}}), \\[2mm] \mathbf{Q}_{i+\frac{1}{2}}^{\mathrm{LW}} = \frac{1}{2}\left(\mathbf{Q}_i^n + \mathbf{Q}_{i+1}^n\right) - \frac{1}{2}\frac{\Delta t}{\Delta x}\left(\mathbf{F}(\mathbf{Q}_{i+1}^n) - \mathbf{F}(\mathbf{Q}_i^n)\right). \end{array}\right\} \tag{134}$$

The FORCE flux is

$$\left.\begin{array}{l} \mathbf{F}_{i+\frac{1}{2}}^{\mathrm{FO}}(\Delta x, \Delta t, \mathbf{Q}_i^n, \mathbf{Q}_{i+1}^n) = \frac{1}{4}\left(\mathbf{F}(\mathbf{Q}_i^n) + 2\mathbf{F}(\mathbf{Q}_{i+\frac{1}{2}}^{\mathrm{LW}}) + \mathbf{F}(\mathbf{Q}_{i+1}^n)\right) \\[3mm] \qquad\qquad - \frac{1}{4}\frac{\Delta x}{\Delta t}\left(\mathbf{Q}_{i+1}^n - \mathbf{Q}_i^n\right). \end{array}\right\} \tag{135}$$

The Lax–Friedrichs and FORCE schemes are first-order accurate and also monotone (for the scalar case). Note that not all first order methods are monotone. The Lax–Wendroff schemes are second-order accurate in space and time but not monotone. All three schemes are called *centred*, or *symmetric*, and are distinct from *upwind schemes*. Schemes of the former class do not use wave propagation information and are very simple to use. Upwind methods, such as the Godunov method, utilise wave propagation information; they are not only more accurate than centred schemes but also more sophisticated.

The finite volume version of the Godunov upwind method is of the form (131), with numerical flux obtained from the solution of the local Riemann problem

$$\left.\begin{array}{ll} \text{PDE:} & \partial_t\mathbf{Q} + \partial_x\mathbf{F}(\mathbf{Q}) = \mathbf{0}, \\[3mm] \text{IC:} & \mathbf{Q}(x,0) = \left\{\begin{array}{ll} \mathbf{Q}_i^n & \text{if } x < 0, \\[3mm] \mathbf{Q}_{i+1}^n & \text{if } x > 0, \end{array}\right. \end{array}\right\} \tag{136}$$

for which the similarity solution is denoted as $\mathbf{Q}_{i+\frac{1}{2}}(x,t)$. The Godunov numerical flux is defined as

$$\mathbf{F}_{i+\frac{1}{2}}^{\text{God}} = \mathbf{F}\left(\mathbf{Q}_{i+\frac{1}{2}}(0)\right). \tag{137}$$

Other definitions of the Godunov numerical flux are also possible. In particular, one can define the Godunov flux in terms of approximate solutions to the Riemann problem, leading to approximate states or directly to approximate fluxes, see [36] for details.

Remarks on Conservative Methods: The important property of monotonicity, valid only for the scalar case, can also be established for conservative methods in terms of the numerical flux. Conservative methods enjoy a number of good properties when it comes to computing solutions to hyperbolic conservations laws. For example, in the presence of shock waves, the use of conservative methods is mandatory. This is supported by numerical and theoretical evidences from the Lax–Wendroff theorem (1960) and the Hou-LeFloch theorem (1994). In particular, it is known that a non-conservative method will compute shock waves with the wrong speed.

2.5 Further Reading

An introduction to numerical methods for hyperbolic conservation laws is found in Chap. 5 of [36]. The books [13, 21] and [19] are particularly recommended. For Riemann solvers to be used in Godunov type methods, see [36]. The library NUMERICA [37] is a collection of source programs for solving hyperbolic equations in one, two and three space dimensions and is suitable for learning and teaching numerical methods. The full library can be downloaded from the website www.ing.unitn.it/toro

3 Two Riemann Solvers: HLLC and EVILIN

Here we discuss two practical approaches for solving the Riemann problem approximately and computing intercell numerical fluxes for Godunov-type methods. The first is called the HLLC solver and was first put forward by Toro et al. [32], see also [33, 34]. HLLC is a modification of the HLL solver proposed by Harten, Lax and van Leer [14]. HLL is based on a two-wave model while HLLC (which stands for Harten, Lax, van Leer and Contact) is based on a three-wave model, resulting in a non-linear complete Riemann solver for the three-dimensional Euler equations, for example. Since HLLC first appeared, several improvements and extensions have been put forward; also, many applications of the method have been published. Useful information is obtained by going to Google and typing *HLLC solver*. The second Riemann solver, we present here, is the EVILIN solver [43]. This is a predictor–corrector approach in which the predictor step evolves in time the initial conditions of the local Riemann problem, then in the corrector step one solves a linearized Riemann problem whose initial conditions are the evolved states. Numerical results for a test problem that illustrates what is distinctive about the two solvers are shown. Relevant references for further reading are indicated at the end of the section.

3.1 The Euler Equations for General Materials

The Euler equations in three space dimensions are

$$\partial_t \mathbf{Q} + \partial_x \mathbf{F}(\mathbf{Q}) + \partial_y \mathbf{G}(\mathbf{Q}) + \partial_z \mathbf{H}(\mathbf{Q}) = \mathbf{0}, \tag{138}$$

with

$$\mathbf{Q} = \begin{bmatrix} \rho \\ \rho u \\ \rho v \\ \rho w \\ E \end{bmatrix}, \mathbf{F} = \begin{bmatrix} \rho u \\ \rho u^2 + p \\ \rho u v \\ \rho u w \\ u(E+p) \end{bmatrix}, \mathbf{G} = \begin{bmatrix} \rho v \\ \rho v u \\ \rho v^2 + p \\ \rho v w \\ v(E+p) \end{bmatrix}, \mathbf{H} = \begin{bmatrix} \rho w \\ \rho w u \\ \rho w v \\ \rho w^2 + p \\ w(E+p) \end{bmatrix}. \tag{139}$$

Here ρ is density; u, v and w are velocity components in the x, y and z directions, respectively; p is pressure and E is total energy given by

$$E = \rho[\frac{1}{2}(u^2 + v^2 + w^2) + e], \tag{140}$$

with e denoting the specific internal energy.

To have a determined system, one requires a closure condition. For general compressible materials, one uses a caloric equation of state relating the variables ρ, p and e. Often one uses other variables, such as the specific volume $1/\rho$ and the entropy s. Here we consider two possible functional relations for a general equilibrium equation of state in terms of the variables ρ, p and e. These are given below, along with the corresponding expressions for the sound speed a in the considered material

$$p = p(\rho, e) \rightarrow a = \sqrt{\frac{p}{\rho^2}p_e + p_\rho}, \quad e = e(\rho, p) \rightarrow a = \sqrt{\frac{p}{\rho^2 e_p} - \frac{e_\rho}{e_p}}, \tag{141}$$

where subscripts denote partial derivatives. We assume the standard convexity condition for the equation of state. For the simple case of ideal gases, one has the familiar equation of state and corresponding sound speed

$$e = \frac{p}{(\gamma - 1)\rho} \quad \rightarrow \quad a = \sqrt{\frac{\gamma p}{\rho}},$$

where γ is the ratio of specific heats. For air under most conditions one takes $\gamma = 1.4$.

3.2 The HLLC Solver

To set the scene, one only needs to consider the conservative scheme

$$\mathbf{Q}_i^{n+1} = \mathbf{Q}_i^n - \frac{\Delta t}{\Delta x} \left(\mathbf{F}_{i+\frac{1}{2}} - \mathbf{F}_{i-\frac{1}{2}} \right). \tag{142}$$

The Godunov upwind finite volume method determines the intercell numerical flux $\mathbf{F}_{i+\frac{1}{2}}$ by solving the Riemann problem for the relevant system of equations with initial conditions $\mathbf{Q}_i^n$ and $\mathbf{Q}_{i+1}^n$.

A Remark on Notation. We shall often use $\mathbf{Q_L}$ to denote $\mathbf{Q}_i^n$ and $\mathbf{QR}$ to denote $\mathbf{Q}_{i+1}^n$.

Consider now the Riemann problem for the three dimensional Euler equations in the direction normal to a finite volume interface. Without loss of generality, we take the x-direction, for which the vectors of conserved variables and fluxes are

$$\mathbf{Q} = \begin{bmatrix} \rho \\ \rho u \\ \rho v \\ \rho w \\ E \end{bmatrix}, \quad \mathbf{F} = \begin{bmatrix} \rho u \\ \rho u^2 + p \\ \rho u v \\ \rho u w \\ u(E + p) \end{bmatrix}. \tag{143}$$

The Riemann problem is the initial-value problem

$$\left. \begin{array}{l} \partial_t \mathbf{Q} + \partial_x \mathbf{F}(\mathbf{Q}) = \mathbf{0}, \\[2mm] \mathbf{Q}(x,0) = \begin{cases} \mathbf{Q_L} \text{ if } x < 0, \\[2mm] \mathbf{Q_R} \text{ if } x > 0. \end{cases} \end{array} \right\} \tag{144}$$

Figure 1 shows the structure of the exact solution of the Riemann problem in terms of the primitive variables $\rho, u, v, w,$ and p for the three-dimensional Euler equations. The eigenvalues are $\lambda_1 = u - a$, $\lambda_2 = \lambda_3 = \lambda_4 = u$, and $\lambda_5 = u + a$, where u is the normal component of velocity and a is the speed of sound. There are five corresponding wave families, of which two are associated with the *acoustic* fields $u - a$ and $u + a$; the middle, coincident, eigenvalues are associated with an entropy wave, a shear wave in the y-direction and a shear wave in the z-direction. The exact solution contains four constant regions: the left and right regions being determined by the initial conditions and two new regions called the *star left* and the *star right* regions, separated by the contact wave. The intermediate wave determines an entropy wave with a jump in density, a shear wave in the y-direction with a jump in v and a shear wave in the z-direction with a jump in w. Pressure and normal velocity are constant throughout the star region, left and right. An approximate Riemann solver must account correctly for these properties of the exact solution.

The fact that three eigenvalues are coincident means that by fitting just one extra wave to the original two-wave HLL solver we are actually able

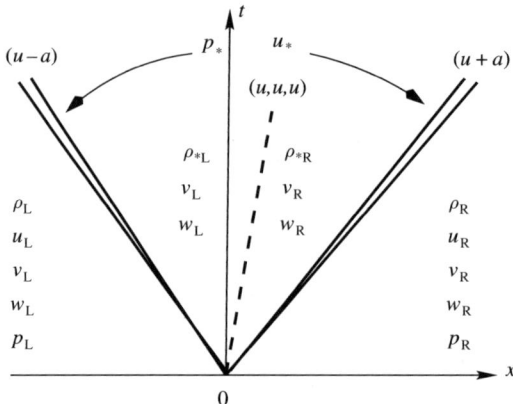

Fig. 1. Structure of the exact solution of the Riemann problem for the three dimensional Euler equations in the x-direction. There are five wave families *associated* with the eigenvalues $u - a$, u (of multiplicity 3) and $u + a$

to account for all five waves in the three-dimensional Euler equations. The HLLC solver assumes a three-wave model for the structure of the solution of the Riemann problem (144) and assumes that wave speed estimates S_L, S_*, S_R for the left, middle and right wave are available. Referring to Fig. 1, the unknown states are $\mathbf{Q}_{*L}$ between the left and middle waves and $\mathbf{Q}_{*R}$ between the middle and right waves. Numerically, we are interested in the corresponding fluxes $\mathbf{F}_{*L}$ and $\mathbf{F}_{*R}$. The HLLC approximate solution for the states, still unknown, satisfies

$$\mathbf{Q}(x,t) = \begin{cases} \mathbf{Q}_L & \text{if } \frac{x}{t} \leq S_L, \\[2mm] \mathbf{Q}_{*L} & \text{if } S_L \leq \frac{x}{t} \leq S_*, \\[2mm] \mathbf{Q}_{*R} & \text{if } S_* \leq \frac{x}{t} \leq S_R, \\[2mm] \mathbf{Q}_R & \text{if } \frac{x}{t} \geq S_R. \end{cases} \tag{145}$$

Integrating the equations over appropriate control volumes yields the following relations

$$\left.\begin{aligned} \mathbf{F}_{*L} &= \mathbf{F}_L + S_L(\mathbf{Q}_{*L} - \mathbf{Q}_L), \\[1mm] \mathbf{F}_{*R} &= \mathbf{F}_{*L} + S_*(\mathbf{Q}_{*R} - \mathbf{Q}_{*L}), \\[1mm] \mathbf{F}_{*R} &= \mathbf{F}_R + S_R(\mathbf{Q}_{*R} - \mathbf{Q}_R). \end{aligned}\right\} \tag{146}$$

These are three equations for the four unknown vectors $\mathbf{Q}_{*L}$, $\mathbf{F}_{*L}$, $\mathbf{Q}_{*R}$ and $\mathbf{F}_{*R}$. The aim is first to find the vectors $\mathbf{Q}_{*L}$ and $\mathbf{Q}_{*R}$ and then the fluxes $\mathbf{F}_{*L}$ and $\mathbf{F}_{*R}$. In order to find a solution, we impose the following conditions on the approximate HLLC solver

$$u_{*L} = u_{*R} = u_*,$$

$$p_{*L} = p_{*R} = p_*,$$

$$v_{*L} = v_L, v_{*R} = v_R, \qquad (147)$$

$$w_{*L} = w_L, w_{*R} = w_R.$$

See Fig. 1. These conditions are in fact satisfied by the exact solution for the Riemann problem (in the normal direction) for the three-dimensional Euler equations.

In addition, by setting

$$S_* = u_* \qquad (148)$$

and using (146), (147) and (148) one can explicitly solve for the vectors $\mathbf{Q}_{*L}$ and $\mathbf{Q}_{*R}$, namely

$$\mathbf{Q}_{*K} = \rho_K \left(\frac{S_K - u_K}{S_K - S_*} \right) \begin{bmatrix} 1 \\ S_* \\ v_K \\ w_K \\ \frac{E_K}{\rho_K} + (S_* - u_K)\left[S_* + \frac{p_K}{\rho_K(S_K - u_K)} \right] \end{bmatrix}, \qquad (149)$$

for $K = L$ and R.

Therefore the fluxes $\mathbf{F}_{*L}$ and $\mathbf{F}_{*R}$ in (146) are completely determined and the HLLC flux reads

$$\mathbf{F}_{i+\frac{1}{2}}^{\text{hllc}} = \begin{cases} \mathbf{F}_L & \text{if} \quad 0 \leq S_L, \\ \mathbf{F}_{*L} = \mathbf{F}_L + S_L(\mathbf{Q}_{*L} - \mathbf{Q}_L) & \text{if } S_L \leq 0 \leq S_*, \\ \mathbf{F}_{*R} = \mathbf{F}_R + S_R(\mathbf{Q}_{*R} - \mathbf{Q}_R) & \text{if } S_* \leq 0 \leq S_R, \\ \mathbf{F}_R & \text{if} \quad 0 \geq S_R. \end{cases} \qquad (150)$$

We note that for any passive scalar q advected with the fluid speed u, the system of equations will include an equation in conservative form

$$(\rho q)_t + (\rho q u)_x = 0. \qquad (151)$$

Such passive scalars may represent concentrations of species in multi-component flow, for example. Note that the tangential velocity components v and w in (143) are special cases of passive scalars.

The new HLLC state due to the passive scalar is given by

$$(\rho q)_{*K} = \rho_K \left(\frac{S_K - u_K}{S_K - S_*} \right) q_K, \qquad (152)$$

for $K = L$ and R.

Wave Speed Estimates

One requires three wave speed estimates: S_L, S_* and S_R. Here we consider a method based on pressure–velocity estimates for the *star region* [36]. Let us suppose that we have estimates p_* and u_* for the pressure and particle velocity in the *star region*. Then we choose the following wave speed estimates

$$S_L = u_L - a_L q_L, \quad S_* = u_*, \quad S_R = u_R + a_R q_R, \tag{153}$$

where

$$q_K = \begin{cases} 1 & \text{if } p_* \le p_K, \\ \left[1 + \frac{\gamma+1}{2\gamma}(p_*/p_K - 1)\right]^{\frac{1}{2}} & \text{if } p_* > p_K. \end{cases} \tag{154}$$

This choice of wave speeds discriminates between shock and rarefaction waves. If the K wave ($K = L$ or R) is a rarefaction then the speed S_K corresponds to the characteristic speed of the head of the rarefaction, which carries the fastest signal. If the wave is a shock wave then the speed corresponds to an approximation to the true shock speed; the wave relations used are exact (for ideal gases) but the pressure ratio across the shock is approximated, because the solution for p_* is an approximation. We propose various schemes to find p_* and u_*.

ALICE: Approximate Linearized Solver Based on the Characteristic Equations. This very simple linearized approximate Riemann solver based on the characteristic form of the equations was put forward in [35]. Approximations for p_* and u_* are

$$\left.\begin{array}{l} p_{ce} = \frac{1}{C_L+C_R}[C_R p_L + C_L p_R + C_L C_R (u_L - u_R)], \\[2mm] u_{ce} = \frac{1}{C_L+C_R}[C_L u_L + C_R u_R + (p_L - p_R)], \end{array}\right\} \tag{155}$$

with $C_L = \rho_L a_L$ and $C_R = \rho_R a_R$.

Two-Rarefaction Riemann Solver. Approximate values for pressure and velocity are given by

$$\left.\begin{array}{l} p_{tr} = \left[\dfrac{a_L + a_R - \frac{\gamma-1}{2}(u_R - u_L)}{a_L/p_L^z + a_R/p_R^z}\right]^{\frac{1}{z}}, \\[4mm] u_{tr} = \dfrac{P_{LR} u_L/a_L + u_R/a_R + \frac{2(P_{LR}-1)}{(\gamma-1)}}{P_{LR}/a_L + 1/a_R}, \end{array}\right\} \tag{156}$$

where

$$P_{LR} = \left(\frac{p_L}{p_R}\right)^z; \quad z = \frac{\gamma-1}{2\gamma}. \tag{157}$$

The Two-Shock Riemann Solver. This is a very robust approximate solver but it does require an estimate p_{*0} for the pressure. This may be obtained from any of the approximate expressions for pressure in (155) or (156). It is then recommended to set $p_0 = max(0, p_{*0})$. The two-shock solutions for pressure and velocity are

$$\left. \begin{aligned} p_{ts} &= \frac{g_L(p_0)p_L + g_R(p_0)p_R - (u_R - u_L)}{g_L(p_0) + g_R(p_0)}, \\[2mm] u_{ts} &= \frac{1}{2}(u_L + u_R) + \frac{1}{2}\left[(p_{ts} - p_R)g_R(p_0) - (p_{ts} - p_L)g_L(p_0)\right], \end{aligned} \right\} \tag{158}$$

where

$$g_K(p_0) = \left[\frac{A_K}{p_0 + B_K}\right]^{\frac{1}{2}}, \tag{159}$$

for $K = $ L and R.

An alternative method assumes only an approximation for the pressure p_*, from which one first computes S_L and S_R, as in (153). Then S_*, an estimate for the middle wave speed, is computed as

$$S_* = \frac{p_R - p_L + \rho_L u_L(S_L - u_L) - \rho_R u_R(S_R - u_R)}{\rho_L(S_L - u_L) - \rho_R(S_R - u_R)}. \tag{160}$$

Summary of HLLC: To find the HLLC numerical flux in (150), one performs the following steps:

- Find pressure–velocity estimates p_* and u_* using any of (155) or (156) or (158);
- Calculate wave speed estimates S_L, S_* and S_R as in (153) and (154); note option (160) for S_*;
- Calculate the states $\mathbf{Q}_{*L}$ or $\mathbf{Q}_{*R}$ in (149), as appropriate; and
- Compute the numerical flux as in (150).

3.3 The EVILIN Riemann Solver

The Riemann problem (144) is solved approximately to obtain the similarity solution $\mathbf{Q}_{LR}(x/t)$ to be used in the computation of the intercell numerical flux as $\mathbf{F}_{i+\frac{1}{2}} = \mathbf{F}(\mathbf{Q}_{LR}(0))$. We study the EVILIN solver [44], a variant of the recently proposed MUSTA approach [41, 43], a multi-stage predictor–corrector scheme in which simple methods are used at each stage. The EVILIN scheme has two stages: a predictor and a corrector, as seen below.

Predictor Step: Data Evolution

The initial conditions in (144) are time evolved as follows:

$$\hat{\mathbf{Q}}_{\mathrm{L}} = \mathbf{Q}_{\mathrm{L}} - \frac{\delta t}{\delta x}[\mathbf{F}_{\mathrm{LR}}^{\mathrm{P}} - \mathbf{F}(\mathbf{Q}_{\mathrm{L}})], \quad \hat{\mathbf{Q}}_{\mathrm{R}} = \mathbf{Q}_{\mathrm{R}} - \frac{\delta t}{\delta x}[\mathbf{F}(\mathbf{Q}_{\mathrm{R}}) - \mathbf{F}_{\mathrm{LR}}^{\mathrm{P}}]. \quad (161)$$

Here $\mathbf{F}_{\mathrm{LR}}^{\mathrm{P}} = \mathbf{F}_{\mathrm{LR}}^{\mathrm{P}}(\mathbf{Q}_{\mathrm{L}}, \mathbf{Q}_{\mathrm{R}})$ is a two-point predictor numerical flux. The computational parameters δx and δt are independent of the mesh parameters Δx and Δt of the scheme (142) and will be specified below.

One possible choice for the predictor is the FORCE flux [38]

$$\left.\begin{array}{l}\mathbf{F}_{i+\frac{1}{2}}^{\mathrm{FO}} = \frac{1}{4}\left[\mathbf{F}(\mathbf{Q}_i^n) + 2\mathbf{F}(\mathbf{Q}_{i+\frac{1}{2}}^{\mathrm{LW}}) + \mathbf{F}(\mathbf{Q}_{i+1}^n) - \frac{\Delta x}{\Delta t}\left(\mathbf{Q}_{i+1}^n - \mathbf{Q}_i^n\right)\right], \\[2mm] \mathbf{Q}_{i+\frac{1}{2}}^{\mathrm{LW}} = \frac{1}{2}[\mathbf{Q}_i^n + \mathbf{Q}_{i+1}^n] - \frac{1}{2}\frac{\Delta t}{\Delta x}[\mathbf{F}(\mathbf{Q}_{i+1}^n) - \mathbf{F}(\mathbf{Q}_i^n)].\end{array}\right\} \quad (162)$$

Another choice is the GFORCE flux [43]

$$\mathbf{F}_{i+\frac{1}{2}}^{\mathrm{GF}} = \Omega\mathbf{F}_{i+\frac{1}{2}}^{\mathrm{LW}} + (1 - \Omega)\mathbf{F}_{i+\frac{1}{2}}^{\mathrm{LF}}, \quad (163)$$

where $\mathbf{F}_{i+\frac{1}{2}}^{\mathrm{LW}} = \mathbf{F}(\mathbf{Q}_{i+\frac{1}{2}}^{\mathrm{LW}})$ is the two-step Lax–Wendroff flux, $\mathbf{F}_{i+\frac{1}{2}}^{\mathrm{LF}}$ is the Lax-Friedrichs flux and

$$\Omega(C) = \frac{1}{1 + C}. \quad (164)$$

C is a prescribed, independent CFL coefficient with $0 < C \leq 1$. We remark that for the linear advection equation $\mathbf{F}_{i+\frac{1}{2}}^{\mathrm{GF}}$ reduces to the Godunov flux if the C is the CFL coefficient of the scheme (142).

Regarding the choice of the parameters δt and δx in the predictor step (161), first note that we can set δx arbitrarily; two obvious choices are $\delta x = 1$ and $\delta x = \Delta x$. Then the *time step* δt is computed exclusively from local wave speed information contained in the two data states $\mathbf{Q}_i^n$, $\mathbf{Q}_{i+1}^n$ and the chosen *CFL coefficient* C, with $0 < C \leq 1$. There is the temptation to call δx, δx and C, *local mesh parameters*, but that terminology would be misleading.

Note also that when applying $\mathbf{F}_{i+\frac{1}{2}}^{\mathrm{GF}}$ in the predictor step (161), the flux depends on the data states $\mathbf{Q}_i^n$ and $\mathbf{Q}_{i+1}^n$ as well as on the parameters δt and δx. Then the GFORCE flux reads

$$\mathbf{F}_{i+\frac{1}{2}}^{\mathrm{GF}} = \Omega\mathbf{F}_{i+\frac{1}{2}}^{\mathrm{LW}}(\mathbf{Q}_i^n, \mathbf{Q}_{i+1}^n, \delta x, \delta t) + (1 - \Omega)\mathbf{F}_{i+\frac{1}{2}}^{\mathrm{LF}}(\mathbf{Q}_i^n, \mathbf{Q}_{i+1}^n, \delta x, \delta t). \quad (165)$$

Corrector Step: Linearized Riemann Solver

In the corrector step, we solve the Riemann problem (144) but with evolved initial conditions $(\hat{\mathbf{Q}}_{\mathrm{L}}, \hat{\mathbf{Q}}_{\mathrm{R}})$ obtained from the predictor step (161). Now we first reformulate the problem in terms of the vector of primitive variables $\mathbf{W} = [\rho, u, v, w, p]^{\mathrm{T}}$ so that the Riemann problem becomes

$$\partial_t \mathbf{W} + \mathbf{B}\partial_x \mathbf{W} = \mathbf{0},$$

$$\hat{\mathbf{W}}(x,0) = \left\{ \begin{array}{l} \hat{\mathbf{W}}_{\mathrm{L}} \text{ if } x < 0, \\ \\ \hat{\mathbf{W}}_{\mathrm{R}} \text{ if } x > 0, \end{array} \right\} \tag{166}$$

with $\hat{\mathbf{W}}_{\mathrm{L}} = \hat{\mathbf{W}}_{\mathrm{L}}(\hat{\mathbf{Q}}_{\mathrm{L}})$ and $\hat{\mathbf{W}}_{\mathrm{R}} = \hat{\mathbf{W}}_{\mathrm{R}}(\hat{\mathbf{Q}}_{\mathrm{R}})$. Then we perform a local linearisation of the system in (166), based on the arithmetic mean

$$\tilde{\mathbf{W}} = \frac{1}{2}(\hat{\mathbf{W}}_{\mathrm{L}} + \hat{\mathbf{W}}_{\mathrm{R}}). \tag{167}$$

A Remark on Notation. Here a quantity $\hat{\eta}$ means an evolved quantity in the predictor step and the quantity $\tilde{\psi}$ means an arithmetic average of corresponding evolved quantities $\hat{\eta}$.

The coefficient matrix of the linearized system is

$$\hat{\mathbf{B}} = \mathbf{B}\left(\tilde{\mathbf{W}}\right). \tag{168}$$

The eigenvalues and eigenvectors of $\hat{\mathbf{B}}$ are denoted by

$$\hat{\lambda}_i = \lambda_i\left(\tilde{\mathbf{W}}\right), \quad \hat{\mathbf{R}}_i = \hat{\mathbf{R}}_i\left(\tilde{\mathbf{W}}\right), \quad \text{for } i = 1, 2, \ldots, m. \tag{169}$$

The similarity solution $\hat{\mathbf{W}}_{\mathrm{LR}}(x/t)$ of (166) is obtained from standard theory of hyperbolic systems with constant coefficients. The jump $\hat{\boldsymbol{\Delta}} \equiv \hat{\mathbf{W}}_{\mathrm{R}} - \hat{\mathbf{W}}_{\mathrm{L}}$ in the initial conditions is given as

$$\hat{\boldsymbol{\Delta}} = [\Delta\hat{\rho}, \Delta\hat{u}, \Delta\hat{v}, \Delta\hat{w}, \Delta\hat{p}]^{\mathrm{T}} = [\hat{\rho}_{\mathrm{R}} - \hat{\rho}_{\mathrm{L}}, \hat{u}_{\mathrm{R}} - \hat{u}_{\mathrm{L}}, \hat{v}_{\mathrm{R}} - \hat{v}_{\mathrm{L}}, \hat{w}_{\mathrm{R}} - \hat{w}_{\mathrm{L}}, \hat{p}_{\mathrm{R}} - \hat{p}_{\mathrm{L}}]^{\mathrm{T}}, \tag{170}$$

which is then projected onto the eigenvectors as

$$\alpha_1\hat{\mathbf{R}}_1 + \alpha_2\hat{\mathbf{R}}_2 + \ldots + \alpha_m\hat{\mathbf{R}}_m = \hat{\boldsymbol{\Delta}}. \tag{171}$$

This is a linear algebraic system for the wave strengths α_i, $i = 1, 2, \ldots, m$, which are the unknowns. For some problems of practical interest, the closed-form solution of the linear system can be easily obtained *by hand*. For more complicated systems, we recommend the use of algebraic manipulators. One may also find the solution numerically using any standard software for linear algebraic systems.

Having found the wave strengths α_i, one knows the solution everywhere in the half plane $t > 0$, $-\infty < x < \infty$. We are interested in the solution at the particular point $x/t = 0$ to determine the sought numerical fluxes $\mathbf{F}_{i+\frac{1}{2}}$. We have the following three options

$$\hat{\mathbf{W}}_{\mathrm{LR}}(0) = \hat{\mathbf{W}}_{\mathrm{L}} + \sum_{\hat{\lambda}_i < 0} \alpha_i \hat{\mathbf{R}}_i, \tag{172}$$

or

$$\hat{\mathbf{W}}_{\mathrm{LR}}(0) = \hat{\mathbf{W}}_{\mathrm{R}} - \sum_{\hat{\lambda}_i > 0} \alpha_i \hat{\mathbf{R}}_i, \tag{173}$$

or

$$\hat{\mathbf{W}}_{\mathrm{LR}}(0) = \frac{1}{2}(\hat{\mathbf{W}}_{\mathrm{L}} + \hat{\mathbf{W}}_{\mathrm{R}}) - \frac{1}{2}\sum_{i=1}^{m} \alpha_i \hat{\mathbf{R}}_i. \tag{174}$$

The sought numerical flux for the finite volume scheme (142) is

$$\mathbf{F}_{i+\frac{1}{2}} = \mathbf{F}(\hat{\mathbf{W}}_{\mathrm{LR}}(0)). \tag{175}$$

Summary of EVILIN: To compute the EVILIN numerical flux, one performs the following steps:

- At each volume interface with data $(\mathbf{Q}_{\mathrm{L}}, \mathbf{Q}_{\mathrm{R}})$, apply the predictor step (161) to evolve $(\mathbf{Q}_{\mathrm{L}}, \mathbf{Q}_{\mathrm{R}})$ to $(\hat{\mathbf{Q}}_{\mathrm{L}}, \hat{\mathbf{Q}}_{\mathrm{R}})$;
- Reformulate the Riemann problem with evolved initial conditions $(\hat{\mathbf{Q}}_{\mathrm{L}}, \hat{\mathbf{Q}}_{\mathrm{R}})$ in terms of a suitable set of variables, such as the primitive variables, for example;
- Linearize the Riemann problem with evolved initial conditions to find the state $\hat{\mathbf{W}}_{\mathrm{LR}}(0)$ at the interface;
- Compute the numerical flux as in (175).

Corrector Step for the Euler Equations

For the three-dimensional Euler equations, the linear algebraic system (171) has solution

$$\left. \begin{aligned} \alpha_1 &= \frac{\Delta\hat{p} - \Delta\hat{u}\tilde{\rho}\tilde{a}}{2\tilde{\rho}\tilde{a}^2}, \\[2mm] \alpha_2 &= \frac{\Delta\hat{\rho}\tilde{a}^2 - \Delta\hat{p}}{\tilde{a}^2}, \\[2mm] \alpha_3 &= \Delta\hat{v}, \\[2mm] \alpha_4 &= \Delta\hat{w}, \\[2mm] \alpha_5 &= \frac{\Delta\hat{p} + \Delta\hat{u}\tilde{\rho}\tilde{a}}{2\tilde{\rho}\tilde{a}^2}. \end{aligned} \right\} \tag{176}$$

The explicit solution in the unknown *star region*, see Fig. 1, is given by

$$
\left.
\begin{aligned}
p^* &= \tfrac{1}{2}(\hat{p}_{\mathrm{L}} + \hat{p}_{\mathrm{R}}) - \tfrac{1}{2}(\hat{u}_{\mathrm{R}} - \hat{u}_{\mathrm{L}})C_1, \\[4pt]
u^* &= \tfrac{1}{2}(\hat{u}_{\mathrm{L}} + \hat{u}_{\mathrm{R}}) - \tfrac{1}{2}(\hat{p}_{\mathrm{R}} - \hat{p}_{\mathrm{L}})/C_1, \\[4pt]
\rho_{\mathrm{L}}^* &= \hat{\rho}_{\mathrm{L}} + (\hat{u}_{\mathrm{L}} - u^*)C_2, \\[4pt]
\rho_{\mathrm{R}}^* &= \hat{\rho}_{\mathrm{R}} + (u^* - \hat{u}_{\mathrm{R}})C_2, \\[4pt]
v_{\mathrm{L}}^* &= \hat{v}_{\mathrm{L}}, \\[4pt]
v_{\mathrm{R}}^* &= \hat{v}_{\mathrm{R}}, \\[4pt]
w_{\mathrm{L}}^* &= \hat{w}_{\mathrm{L}}, \\[4pt]
w_{\mathrm{R}}^* &= \hat{w}_{\mathrm{R}},
\end{aligned}
\right\}
\tag{177}
$$

where
$$
C_1 = \tilde{\rho}\tilde{a}, \quad C_2 = \tilde{\rho}/\tilde{a}. \tag{178}
$$

For the ideal gas case $\tilde{a} = \sqrt{\frac{\gamma\tilde{p}}{\tilde{\rho}}}$. For details on more general equations of state, see [44].

3.4 Sample Numerical Results and Discussion

Here we show numerical results for a simple but representative test problem, that of an isolated, stationary contact discontinuity for the one-dimensional ideal Euler equations, with $\gamma = 1.4$. The problem is solved in a domain $[0, 1]$ with uniform velocity $u = 0$ and uniform pressure $p = 1$ throughout the domain and with two values for density, namely $\rho_{\mathrm{L}} = 1.4$ for $x < \tfrac{1}{2}$ and $\rho_{\mathrm{R}} = 1.0$ for $x > \tfrac{1}{2}$. The exact solution is precisely the initial condition, namely a stationary isolated contact discontinuity. Numerical results are shown in Fig. 2 for HLLC and in Fig. 3 for EVILIN, in which comparison is made against the exact solution (full line) and the numerical solution obtained from the HLL Riemann solver. HLLC and EVILIN reproduce the exact solution while HLL smears the discontinuity to unacceptable levels; moreover, this numerical diffusion process in the HLL solver will continue without limit in time.

Even if this kind of trivial test problems is precisely the dividing line (the contact discontinuity) between *complete* and *incomplete* Riemann solvers. The numerical difficulties highlighted by this test are identical to those encountered in resolving other types of waves associated with intermediate linearly degenerate characteristic fields, such as shear waves, vortices, material interfaces and reaction fronts. The numerical results of HLL show that it is not sufficient to use an *upwind* method or just *some* Riemann solver. Of course centred methods, those that do not use wave propagation information, will be

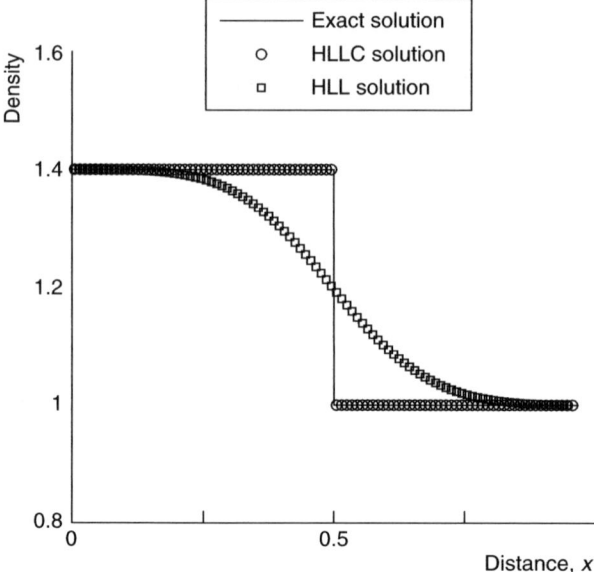

Fig. 2. HLLC solution (*circles*) compared to the exact solution (*line*) and the HLL solution (*squares*) at time 2.0

even more inaccurate than HLL. Capturing shock waves, on the other hand, is not a particularly demanding task nowadays. Incomplete and even good centred methods can give surprisingly satisfactory results. The real challenge is posed by intermediate waves, especially those associated with linearly degenerate fields.

Methods of high order of accuracy may also be brought into the discussion. For smooth solutions, for which high accuracy makes sense, it does not matter

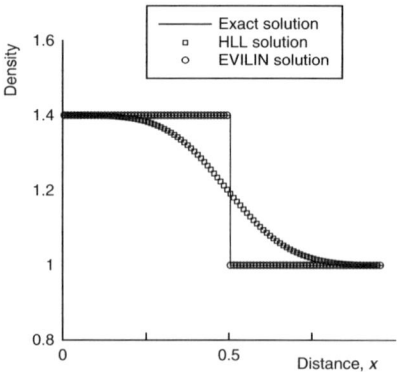

Fig. 3. EVILIN solution (*circles*) compared to the exact solution (*line*) and the HLL solution (*squares*) at time 2.0

which numerical flux is used: a centred scheme such as FORCE, the Rusanov flux (one-wave model), HLL (two-wave model), or any other; all of them should lead to the expected convergence rates. The problem appears for discontinuous solutions, where the resolution of the discontinuities depends very strongly on the numerical flux used. It is often argued that a high-order method can compensate for the incomplete character of the Riemann solver. This is true but only to some extent. Given a linear, slowly moving contact wave, for example, an incomplete Riemann solver will artificially diffuse the wave. Once the linear wave has been diffused, there is no mechanism that will restore back the missing information; moreover, the artificial diffusion mechanism will continue to diffuse the wave in time, without limit. A known exception is a TVD method with the compressive SUPERBEE limiter.

3.5 Further Reading

Background reading on the HLL and HLLC Riemann solvers is found in Chap. 10 of [36]. See the original reference for [14] for HLL and [32, 33] and [34] for HLLC. Further references on HLL are [10], [11] and further references on HLLC are [2], [3]. See [37] for programs for the Euler and shallow water equations in which HLLC is used. Details on EVILIN are found in [44]. Some of the available references on the HLLC solver and related works are also listed here: [1, 16, 22, 23, 24, 25].

4 Non-linear Methods for Scalar Equations

Godunov's theorem states that the only way to have a numerical method of accuracy greater than one and that also avoids the generation of spurious oscillations in the vicinity of large gradients is by constructing non-linear methods, even if applied to linear problems. In this section, we study two classes of non-linear methods: TVD (Total Variation Diminishing) methods and ENO (Essentially Non-Oscillatory) methods. The theory applies to scalar, homogeneous one-dimensional equations. For systems, the scalar theory does not strictly apply but still serves as a useful guide to construct effective numerical methods.

4.1 Monotone Schemes Revisited

Consider a scalar homogeneous conservation law

$$\partial_t q + \partial_x f(q) = 0, \tag{179}$$

where $f(q)$ is the flux function. Conservative numerical methods to solve (179) have the form

$$q_i^{n+1} = q_i^n - \frac{\Delta t}{\Delta x} \left(f_{i+\frac{1}{2}} - f_{i-\frac{1}{2}} \right), \tag{180}$$

where $f_{i+\frac{1}{2}}$ is the numerical flux. Methods to solve (179) can also be written in the more general form

$$q_i^{n+1} = H(q_{i-l}^n, q_{i-l+1}^n, \ldots, q_i^n, \ldots, q_{i+r}^n), \tag{181}$$

where l and r are two non-negative integers that determine the support of the scheme. The so-called *incremental form* is yet another way of expressing a numerical method for (179); these have the form

$$q_i^{n+1} = q_i^n - C_{i-\frac{1}{2}} \Delta q_{i-\frac{1}{2}} + D_{i+\frac{1}{2}} \Delta q_{i+\frac{1}{2}}, \tag{182}$$

where $\Delta q_{i+\frac{1}{2}} = q_{i+1}^n - q_i^n$ and the coefficients $C_{i+\frac{1}{2}}$, $D_{i+\frac{1}{2}}$ are in general assumed to be functions of the data; the schemes are thus non-linear. see [14].

Next we recall monotone methods. A method for (179) written in the form (181) is monotone if

$$\frac{\partial}{\partial q_k^n} H(q_{i-l}^n, q_{i-l+1}^n, \ldots, q_i^n, \ldots, q_{i+r}^n) \geq 0, \quad i - l \leq k \leq i + r. \tag{183}$$

It is also possible to identify monotonicity conditions for a scheme written in conservative form (180). In fact the monotonicity conditions can be applied directly to the numerical flux, as seen in the following theorem.

Theorem. Monotonicity and the flux. *A three-point scheme of the form* (180) *for non-linear conservation laws* (179) *is monotone if*

$$\frac{\partial}{\partial q_i^n} f_{i+\frac{1}{2}}(q_i^n, q_{i+1}^n) \geq 0 \quad \text{and} \quad \frac{\partial}{\partial q_{i+1}^n} f_{i+\frac{1}{2}}(q_i^n, q_{i+1}^n) \leq 0. \tag{184}$$

Consequently, a monotone two-point numerical flux $f_{i+\frac{1}{2}}(q_i^n, q_{i+1}^n)$ *is an increasing (non-decreasing) function of its first argument and a decreasing (non-increasing) function of its second argument.*

Proof. In the scheme (180) we define

$$H(q_{i-1}^n, q_i^n, q_{i+1}^n) \equiv q_i^n - \frac{\Delta t}{\Delta x} \left(f_{i+\frac{1}{2}}(q_i^n, q_{i+1}^n) - f_{i-\frac{1}{2}}(q_{i-1}^n, q_i^n) \right).$$

The required result follows from the following observations

$$\left. \begin{array}{l} \dfrac{\partial H}{\partial q_{i-1}^n} \geq 0 \text{ implies } \dfrac{\partial f_{i-\frac{1}{2}}}{\partial q_{i-1}^n}(q_{i-1}^n, q_i^n) \geq 0, \\[4mm] \dfrac{\partial H}{\partial q_{i+1}^n} \geq 0 \text{ implies } \dfrac{\partial f_{i+\frac{1}{2}}}{\partial q_{i+1}^n}(q_i^n, q_{i+1}^n) \leq 0. \end{array} \right\}$$

Example. the Lax–Friedrichs scheme. For a general non-linear conservation law (179), the Lax–Friedrichs flux is

$$f_{i+\frac{1}{2}}^{LF}(q_i^n, q_{i+1}^n) = \frac{1}{2}\left(f(q_i^n) + f(q_{i+1}^n)\right) - \frac{1}{2}\frac{\Delta x}{\Delta t}\left(q_{i+1}^n - q_i^n\right).$$

Application of conditions (184) to the Lax–Friedrichs flux shows that monotonicity is ensured provided

$$0 \le \frac{\Delta t |\lambda_{\max}|}{\Delta x} \le 1, \tag{185}$$

where $\lambda(q) = \partial f / \partial q$ is the characteristic speed and $|\lambda_{\max}|$ is the maximum in absolute value. That is, provided the CFL stability condition is enforced properly, the Lax–Friedrichs method is monotone when applied to non-linear scalar conservation laws (179).

4.2 TVD Methods

It is possible to construct numerical methods that are better than monotone methods. This can be accomplished by using concepts such as total variation and total variation diminution, as seen below.

Total Variation. Given a mesh function $q^n = \{q_i^n\}$ with q_i^n constant as i tends to $-\infty$ or ∞, the total variation of q^n is defined as

$$TV(q^n) = \sum_{-\infty}^{\infty} |q_{i+1}^n - q_i^n|. \tag{186}$$

TVD Method. A method is said to be TVD if

$$TV(q^{n+1}) \le TV(q^n). \tag{187}$$

Theorem. *(Harten): The set of monotone methods is a subset of the set of TVD methods.*

Remarks on Motone and TVD Methods

- Monotone methods and TVD methods are only defined for scalar equations (179), not for systems.
- The TVD property is a property satisfied by the exact solution of the scalar, linear or non-linear, equation (179); this is not proved here.
- TVD numerical methods attempt to mimic the TVD property of the analytical solution at the discrete level.

- The construction of TVD methods for systems is done on empirical basis; therefore the terminology *TVD method for a system* is not strictly correct.

Theorem. (Harten, 1983): *For any scheme of the form* (182) *to solve* (179), *a sufficient condition for the scheme to be TVD is that the coefficients satisfy*

$$C_{i+\frac{1}{2}} \geq 0, \quad D_{i+\frac{1}{2}} \geq 0, \quad 0 \leq C_{i+\frac{1}{2}} + D_{i+\frac{1}{2}} \leq 1. \tag{188}$$

Proof. See [14], or Chap. 13 of [36].

Remarks on Harten's Theorem

- The coefficients $C_{i+\frac{1}{2}}$ and $D_{i+\frac{1}{2}}$ in Harten's theorem may in general be data dependent.
- The theorem therefore applies to non-linear schemes. This fact can be then used to circumvent Godunov's theorem, which applies to linear schemes only.
- Harten's theorem offers a very useful tool for constructing non-linear schemes of accuracy greater than one.
- Schemes that allow a controlled increase in the total variation have also been constructed. They are usually referred to as Total Variation Bounded (TVB) schemes.

4.3 Flux Limiter Methods

Next we study TVD schemes as for the model linear advection equation

$$\partial_t q + \lambda \partial_x q = 0, \tag{189}$$

with constant coefficient λ.

The class of TVD methods, called flux-limiter methods, have the form

$$f_{i+\frac{1}{2}}^{\text{TVD}} = f_{i+\frac{1}{2}}^{\text{LO}} + \psi_{i+\frac{1}{2}}(f_{i+\frac{1}{2}}^{\text{HO}} - f_{i+\frac{1}{2}}^{\text{LO}}), \tag{190}$$

where $f_{i+\frac{1}{2}}^{\text{LO}}$ is a low-order monotone flux and $f_{i+\frac{1}{2}}^{\text{HO}}$ is a second-order non-monotone flux. Obvious choices are the Godunov's flux for $f_{i+\frac{1}{2}}^{\text{LO}}$ and the Lax–Wendroff flux for $f_{i+\frac{1}{2}}^{\text{HO}}$. The function $\psi_{i+\frac{1}{2}}$ is called *flux limiter* and is constructed on TVD considerations, as seen below. Note that the special value $\psi_{i+\frac{1}{2}} = 0$ gives the monotone flux and $\psi_{i+\frac{1}{2}} = 1$ gives the second order flux.

The Sweby TVD Region. It is possible to derive a region in the r–ψ space, called the TVD region, within which one can select functions, *flux limiters*, that give a TVD method. In order to define this TVD region [31], we first define the parameter r as follows

$$r = \frac{\Delta_{upw}}{\Delta_{loc}} = \begin{cases} \dfrac{q_i^n - q_{i-1}^n}{q_{i+1}^n - q_i^n} = \dfrac{\Delta_{i-\frac{1}{2}}}{\Delta_{i+\frac{1}{2}}} & \text{if } \lambda > 0, \\[3ex] \dfrac{q_{i+2}^n - q_{i+1}^n}{q_{i+1}^n - q_i^n} = \dfrac{\Delta_{i+\frac{3}{2}}}{\Delta_{i+\frac{1}{2}}} & \text{if } \lambda < 0. \end{cases} \tag{191}$$

The TVD region is the shaded portion of Fig. 4, to the right of $\psi = 2r$, above $\psi = 0$ and below $\psi = 2$; also, $\psi = 0$ for $r \le 0$ is part of the TVD region. Any function $\psi(r)$ within this region, called a *flux limiter*, produces a TVD scheme, but not every choice produces a *good* TVD scheme.

Popular choices for flux limiters are the SUPERBEE limiter

$$\psi_{\text{sb}}(r) = \begin{cases} 0 & \text{if} \quad r \le 0, \\ 2r & \text{if} \ 0 \le r \le \frac{1}{2}, \\ 1 & \text{if} \ \frac{1}{2} \le r \le 1, \\ r & \text{if} \ 1 \le r \le 2, \\ 2 & \text{if} \quad r \ge 2, \end{cases} \tag{192}$$

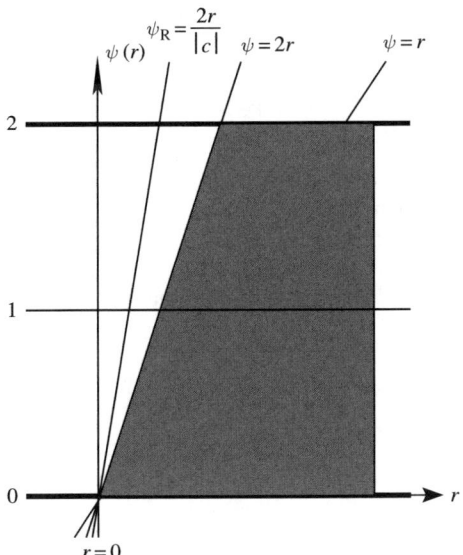

Fig. 4. Sweby's TVD region is the shaded portion to the right of $\psi = 2r$, above $\psi = 0$, below $\psi = 2$ and $\psi = 0$ for $r \le 0$

the VANLEER limiter

$$\psi_{vl}(r) = \begin{cases} 0 & \text{if } r \leq 0, \\ \dfrac{2r}{1+r} & \text{if } r \geq 0, \end{cases} \tag{193}$$

the VANALBADA limiter

$$\psi_{va}(r) = \begin{cases} 0 & \text{if } r \leq 0, \\ \frac{r(1+r)}{1+r^2} & \text{if } r \geq 0, \end{cases} \tag{194}$$

and the MINBEE (or MINMOD) limiter

$$\psi_{mb}(r) = \begin{cases} 0 \ \text{if} & r \leq 0, \\ r \ \text{if } 0 \leq r \leq 1, \\ 1 \ \text{if} & r \geq 1. \end{cases} \tag{195}$$

Remark. Numerical results do depend on the choice of the limiter. They are distinguished by the amount of numerical diffusion added. The least diffusive (also said the most compressive) is SUPERBEE. The most diffusive is MINBEE.

4.4 Reconstruction Methods

A large class of modern finite volume methods of accuracy greater than one use a so-called reconstruction procedure.

Properties of Reconstructions

At any given time level n, one has a set of cell averages $\{q_i^n\}$, which are approximations to integral averages within each cell or finite volume, that is

$$q_i^n \approx \frac{1}{\Delta x} \int\limits_{x_{i-\frac{1}{2}}}^{x_{i+\frac{1}{2}}} q(x, t_n) dx \tag{196}$$

within the cell $[x_{i-\frac{1}{2}}, x_{i+\frac{1}{2}}]$, at time $t = t_n$. These cell averages define a piecewise constant distribution of the solution in the computational domain. To recover the information lost in the averaging, one looks for a set of functions $p_i(x)$ that are defined in a domain that includes $[x_{i-\frac{1}{2}}, x_{i+\frac{1}{2}}]$. Such functions are constructed on the basis of the available cell averages $\{q_i^n\}$. Usually $p_i(x)$ is a polynomial.

Two basic requirements on the reconstruction polynomials are as follows:

1. *The conservation property.* One expects the integral averages of the re-constructed functions to coincide with the original averages, that is

$$q_i^n = \frac{1}{\Delta x} \int\limits_{x_{i-\frac{1}{2}}}^{x_{i+\frac{1}{2}}} p_i(x)dx, \qquad (197)$$

2. *The non-oscillatory property.* There are two different ways of satisfying this requirement: imposing a TVD condition as studied above or imposing an Essentially Non-Oscillatory (ENO) property, as we shall explain.

Example. First-Degree Polynomials. The simplest reconstruction is obtained from first-degree polynomials

$$p_i(x) = q_i^n + (x - x_i)\Delta_i, \qquad (198)$$

where $x_i = \frac{1}{2}(x_{i-\frac{1}{2}} + x_{i+\frac{1}{2}})$ is the cell centre and Δ_i is the slope of the straight line.

Note that the conservation property (197) is satisfied by (192) (verify). To ensure the non-oscillatory property, one selects the slope Δ_i appropriately, for which two criteria are considered, the TVD approach and the ENO approach.

The MUSCL-Hancock Scheme

Here, as an example, we present a specific numerical scheme that uses recon-struction, namely the MUSCL-Hancock scheme. To compute the numerical flux $f_{i+\frac{1}{2}}$ for this method, one performs (see Fig. 5) three steps:

Step (I): Cell boundary values. These are obtained by evaluating the appropriate polynomial at the cell boundaries, namely

$$q_i^L = p_i(x_{i-\frac{1}{2}}) = q_i^n - \frac{1}{2}\Delta x\Delta_i \quad q_i^R = p_i(x_{i+\frac{1}{2}}) = q_i^n + \frac{1}{2}\Delta x\Delta_i. \qquad (199)$$

Step (II): Evolution of cell boundary values.

$$\left.\begin{aligned} \bar{q}_i^L &= q_i^L - \frac{1}{2}\frac{\Delta t}{\Delta x}[f(q_i^R) - f(q_i^L)] = q_i^n - \frac{1}{2}(1+c)\Delta x\Delta_i, \\[2ex] \bar{q}_i^R &= q_i^R - \frac{1}{2}\frac{\Delta t}{\Delta x}[f(q_i^R) - f(q_i^L)] = q_i^n + \frac{1}{2}(1-c)\Delta x\Delta_i. \end{aligned}\right\} \qquad (200)$$

Step (III): The Riemann Problem. To compute the intercell flux $f_{i+\frac{1}{2}}$, one now solves the classical Riemann problem

PDE: $\partial_t q + \partial_x(\lambda q) = 0,$

IC: $q(x,0) = \begin{cases} \bar{q}_i^{\mathrm{R}} = q_i^n + \frac{1}{2}(1-c)\Delta x \Delta_i & \text{if } x/t < \lambda, \\[2mm] \bar{q}_{i+1}^{\mathrm{L}} = q_{i+1}^n - \frac{1}{2}(1+c)\Delta x \Delta_{i+1} & \text{if } x/t > \lambda, \end{cases}$

$$(201)$$

to obtain the similarity solution $q_{i+\frac{1}{2}}(x/t)$. The intercell flux $f_{i+\frac{1}{2}}$ is now computed in *exactly the same way as in the Godunov first-order upwind method*, namely

$$f_{i+\frac{1}{2}} = f(q_{i+\frac{1}{2}}(0)) = \begin{cases} \lambda \bar{q}_i^{\mathrm{R}} = \lambda \left(q_i^n + \frac{1}{2}(1-c)\Delta x \Delta_i\right) & \text{if } \lambda > 0, \\[2mm] \lambda \bar{q}_{i+1}^{\mathrm{L}} = \lambda \left(q_{i+1}^n - \frac{1}{2}(1+c)\Delta x \Delta_{i+1}\right) & \text{if } \lambda < 0. \end{cases}$$

$$(202)$$

The resulting flux is capable of reproducing three well-known second-order schemes: Warming–Beam, Lax–Wendroff and Fromm, depending on the way the slope Δ_i is defined. We have

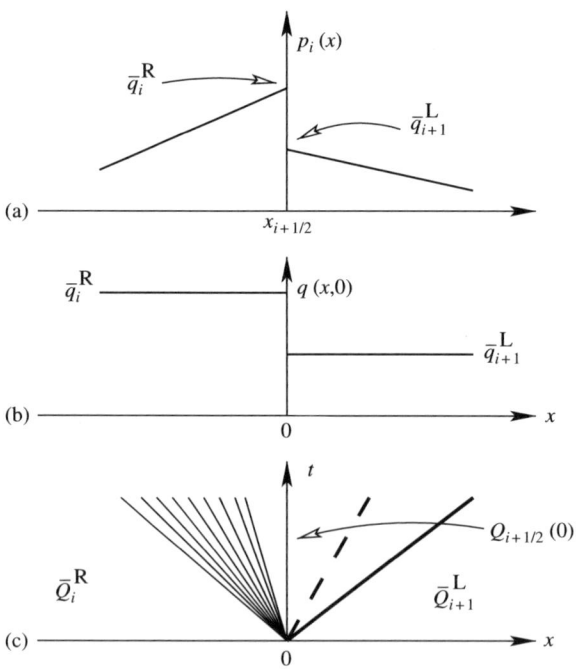

Fig. 5. MUSCL-Hancock schemes. Evolved cell-boundary values (**a**) form the data for a classical Riemann problem (**b**), whose solution (**c**) is used to compute the flux at the interface

$$\Delta_i = \begin{cases} \Delta_{i-\frac{1}{2}} = \dfrac{q_i^n - q_{i-1}^n}{\Delta x} & \text{(Warming–Beam)} \ \lambda > 0, \\[3mm] \Delta_{i+\frac{1}{2}} = \dfrac{q_{i+1}^n - q_i^n}{\Delta x} & \text{(Lax–Wendroff)} \quad \lambda > 0, \\[3mm] \Delta_c = \dfrac{q_{i+1}^n - q_{i-1}^n}{2\Delta x} & \text{(Fromm)}. \end{cases} \qquad (203)$$

Alternatively, instead of solving a Riemann problem and computing the Godunov flux in step (III), one can also use a simple, monotone numerical flux. For example, one can use the FORCE flux. The resulting second-order method has been termed SLIC, see [36]. Application of the FORCE scheme gives the flux

$$f_{i+\frac{1}{2}}^{\text{force}} = f_{i+\frac{1}{2}}^{\text{force}}(\overline{q}_i^{\text{R}}, \overline{q}_{i+1}^{\text{L}}). \qquad (204)$$

But in order to have a non-oscillatory scheme, we must select the slopes in a special manner, as seen below.

TVD Slopes and Slope Limiters

Using TVD criteria one can construct *limited slopes* $\overline{\Delta}_i$ of the form.

$$\overline{\Delta}_i = \overline{\Delta}_i(\Delta_{i-\frac{1}{2}}, \Delta_{i+\frac{1}{2}}). \qquad (205)$$

One possible choice is

$$\overline{\Delta}_i = \begin{cases} max[0, min(\beta\Delta_{i-\frac{1}{2}}, \Delta_{i+\frac{1}{2}}), min(\Delta_{i-\frac{1}{2}}, \beta\Delta_{i+\frac{1}{2}})], \ \Delta_{i+\frac{1}{2}} > 0, \\[3mm] min[0, max(\beta\Delta_{i-\frac{1}{2}}, \Delta_{i+\frac{1}{2}}), max(\Delta_{i-\frac{1}{2}}, \beta\Delta_{i+\frac{1}{2}})], \ \Delta_{i+\frac{1}{2}} < 0, \end{cases} \qquad (206)$$

for particular values of the parameter β.

The value $\beta = 1$ reproduces the MINBEE flux limiter (195), which may also be written as

$$\psi_{mi}(r) = max[0, min(1, r)]. \qquad (207)$$

The value $\beta = 2$ reproduces the SUPERBEE flux limiter (192), which may also be written as

$$\psi_{sb}(r) = max[0, min(2r, 1), min(r, 2)]. \qquad (208)$$

The parameter r is defined in (191).

Alternative Slope Limiters

An alternative approach for obtaining limited slopes is to first define a slope as a linear combination of the slopes $\Delta_{i-\frac{1}{2}}$ and $\Delta_{i+\frac{1}{2}}$, namely

$$\Delta_i^{(c)} = \frac{1}{2}(1+\omega)\Delta_{i-\frac{1}{2}} + \frac{1}{2}(1-\omega)\Delta_{i+\frac{1}{2}}, \quad \omega \in [-1,1]. \tag{209}$$

Then we find a *slope limiter* ξ_i such that

$$\overline{\Delta}_i = \xi_i \Delta_i^{(c)}. \tag{210}$$

This approach leads to a TVD region for $\xi(r)$ given as follows:

$$\xi(r) = 0 \text{ for } r \leq 0, \quad 0 \leq \xi(r) \leq \min\{\xi_L(r), \xi_R(r)\} \text{ for } r > 0. \tag{211}$$

For a simplified version of this region we take

$$\left. \begin{array}{rl} \xi_L(r) &= \dfrac{2r}{1 - \omega + (1+\omega)r}, \\[2mm] \xi_R(r) &= \dfrac{2}{1 - \omega + (1+\omega)r}, \\[2mm] r &= \dfrac{\Delta_{i-\frac{1}{2}}}{\Delta_{i+\frac{1}{2}}}. \end{array} \right\} \tag{212}$$

Figure 6 shows the TVD region for the slope limiter functions. Within this TVD region, one can construct slope limiters that give a TVD method. A slope limiter that is analogous to the SUPERBEE flux limiter is

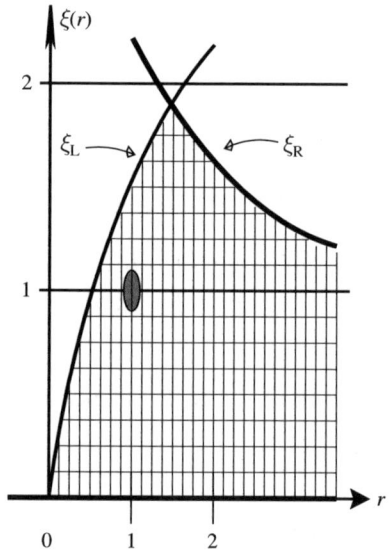

Fig. 6. TVD region for slope limiters

$$\xi_{sb}(r) = \begin{cases} 0 & \text{if } r \le 0, \\ 2r & \text{if } 0 \le r \le \frac{1}{2}, \\ 1 & \text{if } \frac{1}{2} \le r \le 1, \\ \min\left(r, \xi_R(r), 2\right) & \text{if } r \ge 1. \end{cases} \tag{213}$$

A van Leer-type slope limiter is

$$\xi_{vl}(r) = \begin{cases} 0 & \text{if } r \le 0, \\ \min\left(\dfrac{2r}{1+r}, \xi_R(r)\right) & \text{if } r \ge 0. \end{cases} \tag{214}$$

A van Albada-type slope limiter is

$$\xi_{va}(r) = \begin{cases} 0 & \text{if } r \le 0, \\ \min\left(\dfrac{r(1+r)}{1+r^2}, \xi_R(r)\right) & \text{if } r \ge 0. \end{cases} \tag{215}$$

A MINBEE-type slope limiter is

$$\xi_{mb}(r) = \begin{cases} 0 & \text{if } r \le 0, \\ r & \text{if } 0 \le r \le 1, \\ \min\left(1, \xi_R(r)\right) & \text{if } r \ge 1. \end{cases} \tag{216}$$

For details see Chap. 13 of [36].

ENO Slopes

The simplest ENO reconstruction is obtained from first-degree polynomials

$$p_i(x) = q_i^n + (x - x_i)\Delta_i, \tag{217}$$

where $x_i = \frac{1}{2}(x_{i-\frac{1}{2}} + x_{i+\frac{1}{2}})$ is the cell centre and Δ_i is the slope, still to be determined. In the ENO approach, one chooses the slope Δ_i by analysing two potential candidate stencils S_L and S_R and their corresponding polynomials $p_L(x)$ and $p_R(x)$, namely

$$\left. \begin{array}{l} \text{Left stencil:} \quad S_L = \{i-1, i\} \rightarrow p_L(x) = a_L + b_L x, \\ \text{Right stencil:} \quad S_R = \{i, i+1\} \rightarrow p_R(x) = a_R + b_R x, \end{array} \right\} \tag{218}$$

where the coefficients a_L, b_L, a_R and b_R are to be found.

The candidate polynomial for each stencil must satisfy the conservation property. For the left stencil we have

$$\frac{1}{\Delta x} \int_{x_{i-\frac{3}{2}}}^{x_{i-\frac{1}{2}}} p_\mathrm{L}(x)dx = q_{i-1}^n, \quad \frac{1}{\Delta x} \int_{x_{i-\frac{1}{2}}}^{x_{i+\frac{1}{2}}} p_\mathrm{L}(x)dx = q_i^n, \tag{219}$$

yielding

$$\Delta_i \equiv \Delta_{i-\frac{1}{2}} = \frac{q_i^n - q_{i-1}^n}{\Delta x}. \tag{220}$$

Similarly, imposing conservation on $p_\mathrm{R}(x)$

$$\frac{1}{\Delta x} \int_{x_{i-\frac{1}{2}}}^{x_{i+\frac{1}{2}}} p_\mathrm{R}(x)dx = q_i^n, \quad \frac{1}{\Delta x} \int_{x_{i+\frac{1}{2}}}^{x_{i+\frac{3}{2}}} p_\mathrm{R}(x)dx = q_{i+1}^n \tag{221}$$

gives

$$\Delta_i \equiv \Delta_{i+\frac{1}{2}} = \frac{q_{i+1}^n - q_i^n}{\Delta x}. \tag{222}$$

Out of the two possible choices, ENO selects the one with the smallest absolute value, namely

$$\Delta_i = \begin{cases} \Delta_{i-\frac{1}{2}} & \text{if } |\Delta_{i-\frac{1}{2}}| \le |\Delta_{i+\frac{1}{2}}|, \\[2mm] \Delta_{i+\frac{1}{2}} & \text{if } |\Delta_{i-\frac{1}{2}}| > |\Delta_{i+\frac{1}{2}}|. \end{cases} \tag{223}$$

Thus the resulting ENO polynomial is

$$p_i^{\mathrm{ENO}}(x) = \begin{cases} p_\mathrm{L}(x) = q_i^n + (x - x_i)\left(\dfrac{q_i^n - q_{i-1}^n}{\Delta x}\right) & \text{if } |\Delta_{i-\frac{1}{2}}| \le |\Delta_{i+\frac{1}{2}}|, \\[4mm] p_\mathrm{R}(x) = q_i^n + (x - x_i)\left(\dfrac{q_{i+1}^n - q_i^n}{\Delta x}\right) & \text{if } |\Delta_{i-\frac{1}{2}}| > |\Delta_{i+\frac{1}{2}}|. \end{cases} \tag{224}$$

Remarks on Non-linear Schemes

- A modification of the ENO approach, called WENO (for weighted ENO), uses all candidate stencils in the ENO approach to produce a polynomial that is a linear combination of all candidate ENO polynomials. See the work of Jiang and Shu [17].
- The reconstruction approach, ENO or WENO, can be extended to any order of accuracy in one space dimension. WENO has also been extended to multiple space dimensions and on unstructured meshes. For latest developments, see the work of Dumbser et al. [8].

4.5 Further Reading

Most of the material of this section follows Chap. 13 of [36], which is particularly recommended for further reading. The original paper of Harten [14] is a classic on the subject as is the article by Sweby [28]. The textbooks [21] and [19] also contain relevant information on non-linear schemes. In [37] the reader will find a collection of source codes for TVD methods. The paper of Jiang and Shu [17] is a useful source of information on ENO and specially WENO methods. The paper by Dumbser et al. [8] contains some recent developments on non-linear methods of very high order of accuracy for three dimensional problems solved on unstructured meshes.

5 Non-linear Schemes for Hyperbolic Systems

Here we construct non-linear finite volume schemes for non-linear systems of hyperbolic balance laws, with source terms, in one space dimension. The theory is strictly applicable to the scalar homogeneous case only. For systems, the theoretical basis of the scalar case serve only as a useful guide to develop *quasi-non-oscillatory* numerical methods, which in practice turn out to be quite effective. Here we apply the TVD and the ENO criteria to construct second-order finite volume schemes for non-linear systems of conservation laws with source terms.

5.1 Recalling the Finite Volume Method

We consider non-linear systems in one space dimension with source terms

$$\partial_t \mathbf{Q} + \partial_x \mathbf{F}(\mathbf{Q}) = \mathbf{S}(\mathbf{Q}) \tag{225}$$

and finite volume methods to solve (225), which have the form

$$\mathbf{Q}_i^{n+1} = \mathbf{Q}_i^n - \frac{\Delta t}{\Delta x}[\mathbf{F}_{i+\frac{1}{2}} - \mathbf{F}_{i-\frac{1}{2}}] + \Delta t \mathbf{S}_i. \tag{226}$$

Finite volume schemes (226) may be interpreted as resulting from integrating the equations in space and time on the control volume $[x_{i-\frac{1}{2}}, x_{i+\frac{1}{2}}] \times [t_n, t_{n+1}]$ in x–t space. In this manner, (226) is an exact relation in which

- $\mathbf{Q}_i^n$ is the spatial-integral average at time $t = t_n$

$$\mathbf{Q}_i^n = \frac{1}{\Delta x} \int_{x_{i-\frac{1}{2}}}^{x_{i+\frac{1}{2}}} \mathbf{Q}(x, t^n) dx, \tag{227}$$

- $\mathbf{F}_{i+\frac{1}{2}}$ is the time-integral average at the interface $x = x_{i+\frac{1}{2}}$

$$\mathbf{F}_{i+\frac{1}{2}} = \frac{1}{\Delta t} \int_0^{\Delta t} \mathbf{F}(\mathbf{Q}(x_{i+\frac{1}{2}}, t)) dt, \tag{228}$$

- $\mathbf{S}_i$ is the volume-integral average in V

$$\mathbf{S}_i = \frac{1}{\Delta t} \frac{1}{\Delta x} \int_0^{\Delta t} \int_{x_{i-\frac{1}{2}}}^{x_{i+\frac{1}{2}}} \mathbf{S}(\mathbf{Q}_i(x, t)) dx dt. \tag{229}$$

Finite volume numerical methods result from interpreting (226) as an approximate numerical formula to update cell averages $\mathbf{Q}_i^n$, for which one requires a *numerical flux*, still denoted by $\mathbf{F}_{i+\frac{1}{2}}$, and a *numerical source*, still denoted by $\mathbf{S}_i$. In this numerical context, $I_i = [x_{i-\frac{1}{2}}, x_{i+\frac{1}{2}}]$ is the computing cell or just cell; $x_{i-\frac{1}{2}}$ and $x_{i+\frac{1}{2}}$ are the cell interfaces; $\Delta x = x_{i+\frac{1}{2}} - x_{i-\frac{1}{2}}$ is the mesh width or cell width; $x_i = \frac{1}{2}(x_{i-\frac{1}{2}} + x_{i+\frac{1}{2}})$ is the cell centre; and $\Delta t_n = t_{n+1} - t_n$ is the time step or step length. Note that in practice the time step Δt_n varies from time level to time level, but for convenience we often drop the sub-index. To have a method, it is enough to prescribe the *numerical flux* $\mathbf{F}_{i+\frac{1}{2}}$ and the *numerical source* $\mathbf{S}_i$. We study two numerical methods of second order of accuracy, namely MUSCL-Hancock and ADER2, both of them based on polynomial reconstruction.

5.2 Polynomial Reconstruction for Systems

The methods we study here require a reconstruction step to find a reconstruction polynomial vector $\mathbf{P}_i(x)$ for each cell I_i. For systems such as (225), we can perform the reconstruction in two different ways.

Component by Component Reconstructions

For second-order methods with reconstruction in each cell $I_i = [x_{i-\frac{1}{2}}, x_{i+\frac{1}{2}}]$, we construct (or reconstruct) a first-degree polynomial $p_i(x)$ for each component $q = q_k$ of the vector of conserved variables, namely

$$p_i(x) = q_i^n + (x - x_i)\Delta_{i,k}, \tag{230}$$

where $\Delta_{i,k}$ is a slope. Then we apply the reconstruction procedure studied for the scalar case to each component of the vector $\mathbf{Q}$. To choose the slopes $\Delta_{i,k}$, for each component $q = q_k$, one can use either TVD or ENO criteria. Figure 7 shows a single component $p_i(x)$ of the reconstructed vector $\mathbf{P}_i(x)$.

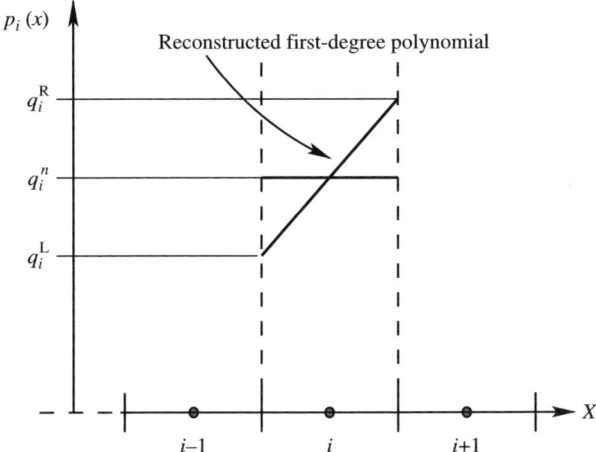

Fig. 7. Illustration of the reconstruction procedure for a single component $p_i(x)$ of the reconstructed vector $\mathbf{P}_i(x)$

Reconstructions in Terms of Characteristic Variables

An alternative to the component-by-component approach is to perform the reconstruction procedure in terms of *characteristic variables*. In this manner, the theory developed for the scalar case is closer to the system case, in some sense. For second-order methods, the advantages of characteristic-variable reconstruction over component-by-component reconstruction are less obvious than for higher order schemes, for which in fact characteristic-variable reconstruction is mandatory.

For each cell $I_i = [x_{i-\frac{1}{2}}, x_{i+\frac{1}{2}}]$, we construct a polynomial $\mathbf{C}_i(x)$ of *characteristic variables* by first transforming from conserved variables to characteristic variables in the usual way. This transformation is based on the *frozen matrix Jacobian matrix* $\mathbf{A}_i^n = \mathbf{A}(\mathbf{Q}_i^n)$ and is carried out not just in cell I_i but in a set of neighbouring cells around I_i with the same matrix $\mathbf{A}_i^n$, namely

$$\mathbf{W}_l = \mathbf{R}_i^{-1}(\mathbf{Q}_l^n), \quad l \in \mathbf{T}_i. \tag{231}$$

$\mathbf{T}_i$ is the set of integers that define the *potential stencils* for the polynomial $\mathbf{C}_i(x)$, for example, $\mathbf{T}_i = \{i-1, i, i+1\}$. $\mathbf{R}_i$ is the matrix whose columns are the right eigenvectors of $\mathbf{A}_i^n$, namely

$$\mathbf{R}_i = \begin{bmatrix} \mathbf{R}_1 \ \mathbf{R}_2 \ \dots \ \mathbf{R}_n \end{bmatrix}, \tag{232}$$

with $\mathbf{R}_k$ the right eigenvector of $\mathbf{A}_i^n$ corresponding to the eigenvalue λ_k, for $k = 1, \dots, m$. $\mathbf{R}_i^{-1}$ is the inverse matrix of $\mathbf{R}_i$.
The vectors $\mathbf{W}_l$ of characteristic variables are denoted as

$$\mathbf{W}_l = \begin{bmatrix} w_{1,l} \\ w_{2,l} \\ \vdots \\ w_{m,l} \end{bmatrix}, \quad l \in \mathbf{T}_i. \tag{233}$$

For each characteristic variable $w_{k,i}$, we build up a reconstructed polynomial, as done for the scalar case.

For second-order schemes, for each component $w_{k,i}$, we write the first-degree polynomial

$$c_i(x) = w_{k,i} + (x - x_i)\Delta_{k,i}. \tag{234}$$

In choosing the slopes $\Delta_{k,i}$, one can apply the TVD criteria studied for the scalar case, in the usual way, for each component of the vector of characteristic variables. The complete reconstruction polynomial $\mathbf{C}_i(x)$ for cell I_i for the characteristic variables is

$$\mathbf{C}_i(x) = \begin{bmatrix} c_1(x) \\ c_2(x) \\ \vdots \\ c_m(x) \end{bmatrix}. \tag{235}$$

When applying the ENO criterion, in terms of characteristic variables, one choose the slopes as follows

$$\Delta_{k,i} = \begin{cases} \dfrac{w_{k,i} - w_{k,i-1}}{\Delta x} & \text{if } |w_{k,i} - w_{k,i-1}| \le |w_{k,i+1} - w_{k,i}|, \\[2mm] \dfrac{w_{k,i+1} - w_{k,i}}{\Delta x} & \text{if } |w_{k,i} - w_{k,i-1}| > |w_{k,i+1} - w_{k,i}|. \end{cases} \tag{236}$$

The method just described gives the complete polynomial $\mathbf{P}_i(x)$ in cell i for the conserved variables, namely

$$\mathbf{P}_i(x) = \mathbf{R}_i \mathbf{C}_i(x). \tag{237}$$

In particular, one may compute cell boundary values, sometimes called *boundary extrapolated values*,

$$\mathbf{Q}_i^{\mathrm{L}} = \mathbf{P}_i(x_{i-\frac{1}{2}}) = \mathbf{R}_i \mathbf{C}_i^{\mathrm{L}}, \quad \mathbf{Q}_i^{\mathrm{R}} = \mathbf{P}_i(x_{i+\frac{1}{2}}) = \mathbf{R}_i \mathbf{C}_i^{\mathrm{R}}. \tag{238}$$

Next we study a scheme that makes use of reconstructions based on first-degree polynomials.

5.3 The MUSCL-Hancock Scheme

In the MUSCL-Hancock approach for a non-linear, homogeneous version of (225), the numerical flux $\mathbf{F}_{i+\frac{1}{2}}$ is computed as follows:

Step (I): Data reconstruction and boundary extrapolated values.
From the reconstruction polynomial vector $\mathbf{P}_i(x)$ in each cell I_i, we obtain
the *boundary extrapolated values* by evaluating $\mathbf{P}_i(x)$ at the boundaries
$x = x_{i-\frac{1}{2}}$ and $x = x_{i+\frac{1}{2}}$, namely

$$\mathbf{Q}_i^{\mathrm{L}} = \mathbf{P}_i(x_{i-\frac{1}{2}}) \quad \mathbf{Q}_i^{\mathrm{R}} = \mathbf{P}_i(x_{i+\frac{1}{2}}). \tag{239}$$

Step (II): Evolution. For each cell I_i, the boundary extrapolated values
$\mathbf{Q}_i^{\mathrm{L}}$ and $\mathbf{Q}_i^{\mathrm{R}}$ in (239) are *evolved* by a time $\frac{1}{2}\Delta t$

$$\left.\begin{aligned}
\overline{\mathbf{Q}}_i^{\mathrm{L}} &= \mathbf{Q}_i^{\mathrm{L}} - \frac{1}{2}\frac{\Delta t}{\Delta x}[\mathbf{F}(\mathbf{Q}_i^{\mathrm{R}}) - \mathbf{F}(\mathbf{Q}_i^{\mathrm{L}})], \\[2mm]
\overline{\mathbf{Q}}_i^{\mathrm{R}} &= \mathbf{Q}_i^{\mathrm{R}} - \frac{1}{2}\frac{\Delta t}{\Delta x}[\mathbf{F}(\mathbf{Q}_i^{\mathrm{R}}) - \mathbf{F}(\mathbf{Q}_i^{\mathrm{L}})].
\end{aligned}\right\} \tag{240}$$

Note that this evolution step is entirely contained in each cell I_i, as the
intercell fluxes are evaluated at the boundary extrapolated values of each
cell. At each intercell position $x_{i+\frac{1}{2}}$, there are two fluxes, namely $\mathbf{F}(\mathbf{Q}_i^{\mathrm{R}})$
and $\mathbf{F}(\mathbf{Q}_{i+1}^{\mathrm{L}})$, which are in general distinct. This does not really affect
the *conservative* character of the overall method, as this step is only an
intermediate step; the intercell flux $\mathbf{F}_{i+\frac{1}{2}}$ to be used in (226) is yet to be
evaluated.

Step (III): The Riemann problem. To compute the intercell flux $\mathbf{F}_{i+\frac{1}{2}}$,
one now solves the classical Riemann problem with data

$$\mathbf{Q}_{\mathrm{L}} \equiv \overline{\mathbf{Q}}_i^{\mathrm{R}} \quad \mathbf{Q}_{\mathrm{R}} \equiv \overline{\mathbf{Q}}_{i+1}^{\mathrm{L}} \tag{241}$$

to obtain the similarity solution $\overline{\mathbf{Q}}_{i+\frac{1}{2}}(x/t)$. The intercell flux $\mathbf{F}_{i+\frac{1}{2}}$ is now
computed in exactly the same way as in the Godunov first-order upwind
method, namely

$$\mathbf{F}_{i+\frac{1}{2}} = \mathbf{F}(\overline{\mathbf{Q}}_{i+\frac{1}{2}}(0)). \tag{242}$$

Here $\overline{\mathbf{Q}}_{i+\frac{1}{2}}(0)$ denotes the value of $\overline{\mathbf{Q}}_{i+\frac{1}{2}}(x/t)$ at $x/t = 0$, i.e. the value of
$\overline{\mathbf{Q}}_{i+\frac{1}{2}}(x/t)$ along the t-axis. Figure 8 illustrates the steps for the MUSCL-
Hancock scheme.
At this stage, one can choose any method by solving the Riemann problem approximately or exactly. Many approximate Riemann solvers give
directly approximations for the flux.

Remark. **centred flux.** Alternatively, instead of solving a Riemann problem
in step (III), one can also use a simple but monotone (checked to be monotone for the scalar case) numerical flux. For example, one can use the FORCE
flux. The resulting second-order method has been termed the SLIC scheme.
Application of the FORCE scheme gives the flux

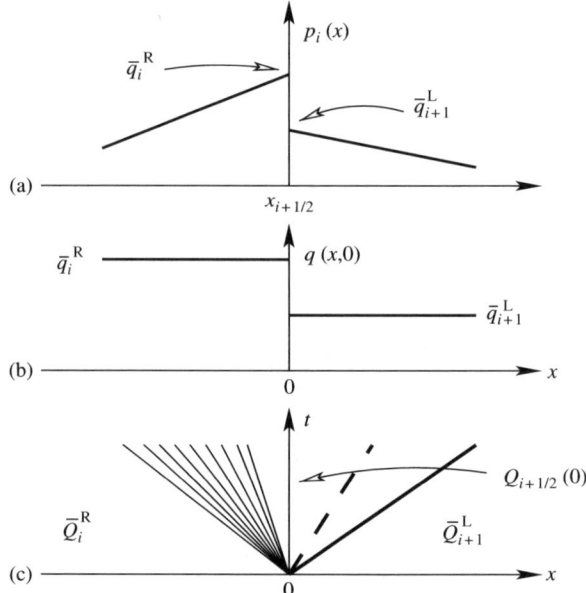

Fig. 8. MUSCL-Hancock schemes. Evolved boundary extrapolated values (**a**) form the data for a classical Riemann problem (**b**), whose solution (**c**) is used to compute the flux at the interface

$$\mathbf{F}^{\text{force}}_{i+\frac{1}{2}} = \mathbf{F}^{\text{force}}_{i+\frac{1}{2}}(\overline{\mathbf{Q}}^{\text{R}}_i, \overline{\mathbf{Q}}^{\text{L}}_{i+1}). \tag{243}$$

So far the scheme has been described for the homogeneous version of the non-linear system (225).

5.4 ADER2 for Non-linear Systems with Source Terms

Here we apply the ADER approach to construct second-order methods to solve non-linear systems of $m \times m$ hyperbolic equations with source terms (225) as solved by finite volume scheme of the type (226). In what follows, we describe the ADER2 scheme for compting the numerical flux $\mathbf{F}_{i+\frac{1}{2}}$ and the numerical source $\mathbf{S}_i$.

The Numerical Flux

To compute the numerical flux $\mathbf{F}_{i+\frac{1}{2}}$, we solve the Derivative Riemann Problem (also called generalized Riemann problem of high-order Riemann problem)

$$\left. \begin{array}{l} \text{PDEs: } \partial_t \mathbf{Q} + \partial_x \mathbf{F}(\mathbf{Q}) = \mathbf{S}(\mathbf{Q}), \\[2mm] \text{IC: } \mathbf{Q}(x,0) = \begin{cases} \mathbf{P}_i(x) & \text{if } x < 0, \\[2mm] \mathbf{P}_{i+1}(x) & \text{if } x > 0, \end{cases} \end{array} \right\} \tag{244}$$

where $\mathbf{P}_i(x)$ is a vector whose components are reconstructed polynomials of degree one. The solution of (244) has the form

$$\mathbf{Q}(0,\tau) = \mathbf{Q}(0,0_+) + \tau\partial_t\mathbf{Q}(0,0_+). \tag{245}$$

The two terms in the expansion are computed as follows:

Step (I): The leading term. To compute the leading term, one solves the classical Riemann problem (piece-wise constant data)

$$\text{PDE:} \quad \partial_t\mathbf{Q} + \partial_x\mathbf{F}(\mathbf{Q}) = \mathbf{0},$$

$$\text{IC:} \quad \mathbf{Q}(x,0) = \left\{ \begin{array}{ll} \mathbf{P}_i(0) & \text{if } x < 0, \\ \\ \mathbf{P}_{i+1}(0) & \text{if } x > 0, \end{array} \right\} \tag{246}$$

in which the initial condition consists of the two boundary extrapolated values from left and right, that is $\mathbf{P}_i(0)$ from the left and $\mathbf{P}_{i+1}(0)$ from the right. Figure 9 illustrates the steps to compute the leading term of the expansion. Denoting the solution of (246) by $\mathbf{D}_{i+\frac{1}{2}}^{(0)}(x/t)$, the leading term is given by

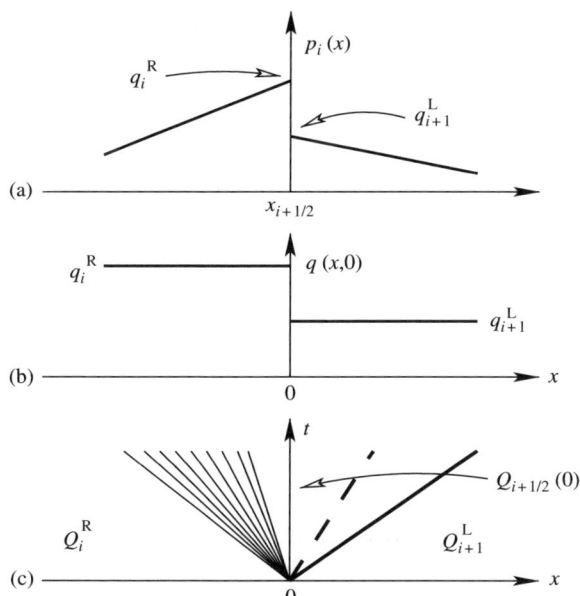

Fig. 9. ADER2 scheme. Boundary extrapolated values from data reconstruction (**a**) form the initial data for a classical Riemann problem (**b**), whose solution (**c**) evaluated at the interface gives the leading term for the ADER2 flux

$$\mathbf{Q}(0,0_+) = \mathbf{D}^{(0)}_{i+\frac{1}{2}}(0). \tag{247}$$

One can use the exact Riemann solver, or any of the approximations available, to compute the leading term.

Step (II): The higher order term. To compute the higher order term in (245), we do the following:

1. **The Cauchy–Kowalewski procedure.** One uses the PDE in (225) to express the time derivative in terms of the space derivative and the source term,

$$\partial_t \mathbf{Q}(x,t) = -\mathbf{A}(\mathbf{Q})\partial_x \mathbf{Q} + \mathbf{S}(\mathbf{Q}), \tag{248}$$

where $\mathbf{A}(\mathbf{Q})$ is the Jacobian matrix.

2. **Evolution equations for the space derivative.** To determine the space derivatives in (248), we first construct an evolution equation for the spatial derivatives $\partial_x \mathbf{Q}(x,t)$, namely

$$\partial_t(\partial_x \mathbf{Q}) + \mathbf{A}(\mathbf{Q})\partial_x(\partial_x \mathbf{Q}) = \mathbf{H}. \tag{249}$$

3. **Riemann problem for the spatial derivatives.** We simplify (249) by neglecting the source term $\mathbf{H}$ and linearising the equations by introducing the constant coefficient matrix $\mathbf{A}^{(0)} = \mathbf{A}(\mathbf{Q}(0,0_+))$. We obtain the linear system with constant coefficients

$$\partial_t(\partial_x \mathbf{Q}) + \mathbf{A}^{(0)}\partial_x(\partial_x \mathbf{Q}) = \mathbf{0}. \tag{250}$$

Then we pose the classical homogeneous Riemann problem for the spatial derivative of the vector $\mathbf{Q}$:

$$\left.\begin{array}{ll} \text{PDEs:} \quad \partial_t(\partial_x \mathbf{Q}) + \mathbf{A}^{(0)}\partial_x(\partial_x \mathbf{Q}) = \mathbf{0} \\[2mm] \text{IC:} \quad \partial_x \mathbf{Q}(x,0) = \begin{cases} \mathbf{\Delta}_i \equiv \mathbf{P}'_i(x) & \text{if } x < 0, \\[2mm] \mathbf{\Delta}_{i+1} \equiv \mathbf{P}'_{i+1}(x) & \text{if } x > 0. \end{cases} \end{array}\right\} \tag{251}$$

We solve this linear Riemann problem with constant coefficients to obtain the similarity solution $\mathbf{D}^{(1)}_{i+\frac{1}{2}}(x/t)$. Then we take $\partial_x \mathbf{Q}(0,0_+) = \mathbf{D}^{(1)}_{i+\frac{1}{2}}(0)$.

Step (III): The solution and the numerical flux. The sought complete solution is

$$\mathbf{Q}(0,\tau) = \mathbf{Q}(0,0_+) + \tau[-\mathbf{A}^{(0)}\mathbf{D}^{(1)}_{i+\frac{1}{2}}(0) + \mathbf{S}(\mathbf{Q}(0,0_+))]. \tag{252}$$

Finally, according to (228), the numerical flux is obtained from

$$\mathbf{F}_{i+\frac{1}{2}} = \frac{1}{\Delta t} \int_0^{\Delta t} \mathbf{F}\left(\mathbf{Q}(0,0_+) + \tau[-\mathbf{A}^{(0)}\mathbf{D}^{(1)}_{i+\frac{1}{2}}(0) + \mathbf{S}(\mathbf{Q}(0,0_+))]\right) d\tau. \quad (253)$$

Integration to second order gives

$$\mathbf{F}_{i+\frac{1}{2}} = \mathbf{F}\left(\mathbf{Q}(0,0_+) + \frac{1}{2}\Delta t[-\mathbf{A}^{(0)}\mathbf{D}^{(1)}_{i+\frac{1}{2}}(0) + \mathbf{S}(\mathbf{Q}(0,0_+))]\right). \quad (254)$$

Note that the numerical flux for the inhomogeneous system (225) depends on the source term. This is obviously not sufficient to complete the scheme; we still need the numerical source.

Numerical Source

To compute the numerical source $\mathbf{S}_i$ in the finite volume scheme (226), we need to select the function $\mathbf{Q}_i(x,t)$ in (229) and an integration scheme. For a second order method, we may apply the mid-point integration rule in space and time, so that one only requires the single value $\mathbf{Q}_i(x_i, \frac{1}{2}\Delta t)$. At time $t = 0$, within the cell i we have the reconstructed polynomial $\mathbf{P}_i(x)$. Applying the Cauchy–Kowalewski method at the point x_i gives

$$\mathbf{Q}(x_i, \tau) = \mathbf{Q}_i^n + \tau \partial_t \mathbf{Q}(x_i, \tau) = \mathbf{Q}_i^n + [-\mathbf{A}(\mathbf{Q}_i^n)\Delta_i + \mathbf{S}(\mathbf{Q}_i^n)]\tau, \quad (255)$$

so that the numerical source, to second order, is

$$\mathbf{S}_i = \mathbf{S}\left(\mathbf{Q}_i^n + \frac{1}{2}\Delta t[-\mathbf{A}(\mathbf{Q}_i^n)\Delta_i + \mathbf{S}(\mathbf{Q}_i^n)]\right). \quad (256)$$

5.5 Source Terms to Second-Order

It is not a trivial matter to construct a numerical scheme for systems with source terms, even of second-order accuracy, that actually preserves second-order for the full scheme. Here we implement and test three second-order methods to deal with hyperbolic equations with source terms. We do so in terms of the model advection–reaction equation

$$\partial_t q + \lambda \partial_x q = \beta q, \quad (257)$$

with a source term $s(q) = \beta q$, where λ and β are two constants.

ADER2

The ADER approach has a very natural way of dealing with source terms to any desired accuracy. Now for the model equation (257), for $\lambda > 0$, application of the ADER2 method gives the numerical flux

$$f_{i+\frac{1}{2}} = \lambda \left[q_i^n + \frac{1}{2}(1-c)\Delta x \Delta_i + \frac{1}{2}r(q_i^n + \frac{1}{2}\Delta x \Delta_i) \right], \tag{258}$$

where

$$c = \frac{\lambda \Delta t}{\Delta x}, \quad r = \Delta t \beta \tag{259}$$

are the CFL number and a *reaction number*, both dimensionless quantities. Note that the flux depends on the source term.

The numerical source, for any value of λ, is

$$s_i = \beta \left[(1 + \frac{1}{2}r)q_i^n - \frac{1}{2}c\Delta x \Delta_i \right]. \tag{260}$$

MUSCL-Hancock

Here we emulate the ADER2 scheme for advection and reaction terms to treat source terms to second-order of accuracy in the MUSCL-Hancock scheme.

To compute the numerical flux, we include the source term in the data evolution step as follows

$$\left. \begin{aligned} \bar{q}_i^L &= q_i^L - \frac{1}{2}\frac{\Delta t}{\Delta x}[f(q_i^R) - f(q_i^L)] + \frac{1}{2}\Delta t(\beta q_i^L), \\ \bar{q}_i^R &= q_i^R - \frac{1}{2}\frac{\Delta t}{\Delta x}[f(q_i^R) - f(q_i^L)] + \frac{1}{2}\Delta t(\beta q_i^R). \end{aligned} \right\} \tag{261}$$

For $\lambda > 0$, simple manipulations give the numerical flux as

$$f_{i+\frac{1}{2}} = \lambda[q_i^n + \frac{1}{2}(1-c)\Delta x \Delta_i + \frac{1}{2}r(q_i^n + \frac{1}{2}\Delta x \Delta_i)]. \tag{262}$$

The numerical source in the MUSCL-Hancock scheme, by construction, is taken to be identical to that of the ADER2 scheme.

WAF

We include here the WAF scheme; see [36] for a full description of the scheme for homogeneous systems. For the WAF method we perform the following steps:

- Evolve cell averages q_i^n and q_{i+1}^n to

$$\left. \begin{aligned} \hat{q}_i^n &= q_i^n + \frac{1}{2}\Delta t \beta q_i^n &= (1 + \frac{1}{2}r)q_i^n, \\ \hat{q}_{i+1}^n &= q_{i+1}^n + \frac{1}{2}\Delta t \beta q_{i+1}^n &= (1 + \frac{1}{2}r)q_{i+1}^n. \end{aligned} \right\} \tag{263}$$

- Solve the classical (piece-wise constant data) Riemann problem with evolved data $\hat{q}_i^n$ and $\hat{q}_{i+1}^n$ to compute the usual WAF flux

$$f_{i+\frac{1}{2}} = \lambda[\frac{1}{2}(1+c)(1 + \frac{1}{2}r)q_i^n + \frac{1}{2}(1-c)(1 + \frac{1}{2}r)q_{i+1}^n]. \tag{264}$$

- Compute the numerical source, as in the ADER2 method,

$$s_i = \beta \left[(1 + \frac{1}{2}r)q_i^n - \frac{1}{2}c\Delta x \Delta_i \right], \tag{265}$$

where, for the WAF method, one needs the additional computation of a slope. We take

$$\Delta_i = \frac{q_{i+\frac{1}{2}} - q_{i-\frac{1}{2}}}{\Delta x}, \tag{266}$$

with

$$\left. \begin{array}{l} q_{i-\frac{1}{2}} = \frac{1}{2}(1 + c)q_{i-1}^n + \frac{1}{2}(1 - c)q_i^n, \\[2mm] q_{i+\frac{1}{2}} = \frac{1}{2}(1 + c)q_i^n + \frac{1}{2}(1 - c)q_{i+1}^n. \end{array} \right\} \tag{267}$$

Other ways of computing the slope Δ_i in the WAF numerical source are also possible.

A Test Problem with a Source Term

We solve the linear advection–reaction equation (257) in the domain $[-1, 1]$, with initial condition

$$q(x, 0) = q^{(0)}(x) = \sin(\pi x) \tag{268}$$

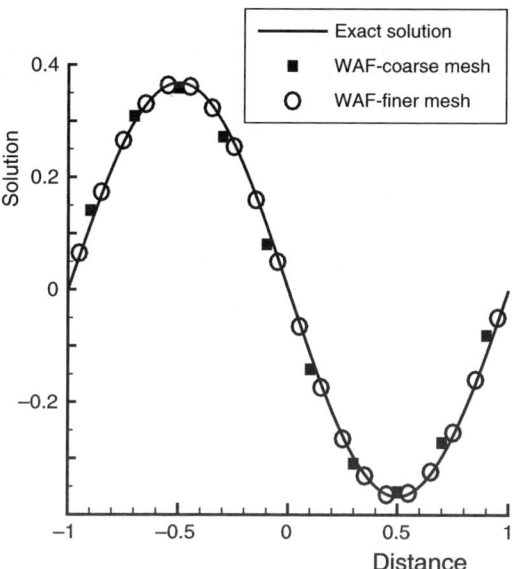

Fig. 10. Model linear advection-reaction equation. Exact and numerical (WAF) solutions

and periodic boundary conditions. The exact solution for any initial condition $q_0(x)$ is

$$q(x, t) = q^{(0)}(x - \lambda t)e^{\beta t}. \tag{269}$$

For the numerical experiments, we take $\lambda = 1$ and $\beta = -1$. Figure 10 shows a comparison between the exact solution (line) and the WAF numerical solutions for two meshes. Table 1 shows the convergence rates measured in two norms. As can be seen, the expected second-order of accuracy is achieved.

Table 1. Advection and source term. Convergence rates for the WAF method in two norms, for a sequence of four meshes ($M = 10, 20, 40, 80$)

M	L_∞-error	L_∞-order	L_1-error	L_1-order
10	3.28×10^{-2}		4.24×10^{-2}	
20	7.86×10^{-3}	2.06	1.00×10^{-2}	2.08
40	1.82×10^{-3}	2.11	2.33×10^{-3}	2.10
80	4.51×10^{-4}	2.02	5.74×10^{-4}	2.02

5.6 Advanced Methods

In these five Sections I have presented the basics on numerical methods for solving hyperbolic equations. I introduced finite difference methods in order to analyse some of the main properties of the schemes. Most of the material is however on finite volume methods. The treatment has also been restricted to one-dimensional non-nonlinear systems with (non-stiff) source terms. Two non-linear methods of second-order of accuracy have been studied, the MUSCL-Hancock method and the ADER2 method. Both schemes can be extended to solve conservation laws in multiple space dimensions on structured and unstructured meshes. For further reading on the subject, we recommend the textbooks [36] and [21].

Regarding the advanced numerical methods of the future to solve non-linear hyperbolic systems of balance laws, my wish list would require numerical schemes to be able to

- treat stiff source terms;
- deal with multiple space dimensions;
- deal with complicated geometries;
- use unstructured meshes;
- have no theoretical accuracy barrier, neither in space nor in time;
- have the potential to accommodate the inclusion of other physical effects, such as dissipative and dispersive effects.

With the pioneering work of Harten and collaborators, followed by more recent developments, it seems to me as if we might be able to achieve the above ambitious goals in the near future and to have potent computational methods available for advanced scientific computing.

The ADER approach [39] allows the construction of schemes of arbitrary order of accuracy in both space and time. It is a generalisation to arbitrary order of accuracy of the so-called Generalized Riemann Problem method (or GRP method) of Ben-Artzi and Falcoviz [4], which is a second-order scheme. See the original paper of Toro and collaborators [39] for preliminary results, where schemes were formulated for linear equations in one, two and three space dimensions on cartesian meshes.

The ADER approach is closely related to the high-order method of Harten and collaborators [15]. See [6] for a comparative study. It could be said that the ADER method is a generalization to arbitrary order of accuracy of the second-order GRP scheme of Ben-Artzi and Falcovitz, and that the scheme of Harten and collaborators is a generalisation of the second-order MUSCL-Hancock scheme. Both approaches can be unified by considering the high-order Riemann problem, that is the Cauchy problem in which the initial conditions are, for example, high-order polynomials in space, and the solution at the interface is a high-order polynomial in time.

Extensions of the ADER method to non-linear systems with source terms were made possible by solving the high-order Riemann problem to any accuracy; see [30, 40, 43]. The generalisation to multidimensional non-linear problems on regular meshes is reported in [31] and [42]. The methodology has been extended in a variety of ways and applied to a variety of physical problems. See, for example, [7, 18, 26, 29].

The recent paper of Dumbser and collaborators [8] reports on the extension of the ADER approach to solve the three-dimensional Euler equations on unstructured meshes, using schemes of very high order of accuracy in both space and time. The approach has also been extended to solve hyperbolic balance laws with stiff source terms, reconciling (apparently for the first time) three conflicting items: stiff source terms, explicit methods and high-order of accuracy in space and time; see the work of Dumbser and collaborators [9]. The ADER approach has also been extended to solve parabolic equations. See the works of Toro and Hidalgo [45], who solved non-linear diffusion–reaction equations with schemes of up to 10th order of accuracy. See also the recent paper by Gasnner and collaborators [12].

References

1. Ball G.J., A free-Lagrange method for unsteady compressible flow: Simulation of a confined cylindrical blast wave. JShWa **5**(5), 311–325, February (1996)
2. Batten, P., Clarke, N., Lambert, C., & Causon, D.M., On the choice of wave speeds for the HLLC Riemann solver. SIAM JSSC, **18**, 1553–1570 (1997)
3. Batten, P., Leschziner, M.A., & Goldberg, U.C., Average state Jacobians and implicit methods for compressible viscous and turbulent flows. JCP, **137**, 38–78, (1997)

4. Ben-Artzi, M., & Falcovitz, J., A second-order Godunov-type scheme for compressible fluid dynamics. JCP, **55**, 1–32 (1984)
5. Bressan, A., Hyperbolic Systems of Conservation Laws, Oxford University Press, Oxford (2000)
6. Castro, C.E., & Toro, E.F., Alternative solvers for the derivative Riemann problem for hyperbolic balance laws. Isaac Newton Institute for Mathematical Sciences, University of Cambridge, UK, Preprint Series, NI06053-NPA (2007)
7. Dumbser, M., & Käser, M., An arbitrary high order discontinuous Galerkin method for elastic waves on unstructured meshes II: The three-dimensional isotropic case. GJI, **167**(1), 319–336 (2006)
8. Dumbser, M., Käser, M., Titarev, V.A., & Toro, E.F., Quadrature-free non-oscillatory finite volume schemes on unstructured meshes for non-linear hyperbolic systems. JCP, **221**(2), 693–723 (2007)
9. Dumbser, M., Enaux, C., & Toro, E.F., Explicit finite volume schemes of arbitrary high order of accuracy for hyperbolic systems with stiff source terms. Isaac Newton Institute for Mathematical Sciences, University of Cambridge, UK, Preprint Series, NI07007-NPA (2007)
10. Einfeldt, B., On Godunov-type methods for gas dynamics. SIAMJNA, **25**(2), 294–318 (1988)
11. Einfeldt, B., Munz, C.D., Roe, P.L., & Sjoegreen, B., On Godunov-type methods near low densities. JCP, **92**, 273–291 (1991)
12. Gassner, G., Lörcher, F., & Munz, C.D., A contribution to the construction of difussion fluxes for finite volume schemes and discontinuous Galerkin schemes. JCP, **224**(2), 1049–1063 (2007)
13. Godlewski, E., & Raviart, P.A., Numerical Approximation of Hyperbolic Systems of Conservation Laws, Springer, Berlin (1996)
14. Harten, A., Lax, P., & van Leer, B., On upstream differencing and Godunov-type schemes for hyperbolic conservation laws. SIAMR, **25**(1), 35–61 (1983)
15. Harten, A., Engquist, B., Osher, S., & Chakravarthy, S.R., Uniformly high order accuracy essentially non-oscillatory schemes III. JCP, **71**, 231–303 (1987)
16. Honkkila, V., Janhunen, P., HLLC solver for ideal relativistic MHD. JCP, **223**(2), 643–656, May (2007)
17. Jiang, G.S., & Shu, C.W., Efficient implementation of Weighted ENO schemes. JCP, **126**, 202–212 (1996)
18. Käser, M., & Dumbser, M., An arbitrary high order discontinuous Galerkin method for elastic waves on unstructured meshes I: The two-dimensional isotropic case with external source terms. GJI, **166**(2), 855–877 (2006)
19. Laney, C.B., Computational Gas Dynamics. Cambridge University Press, Cambridge (1998)
20. LeFloch, P.G., Hyperbolic Systems of Conservation Laws: The Theory of Classical and Nonclassical Shock Waves, Lectures in Mathematics, ETH Zürich, Birkhäuser (2002)
21. LeVeque, R.J., Finite Volume Methods for Hyperbolic Problems, Cambridge Univesity Press, Cambridge (2002)
22. Li, S., An HLLC Riemann solver for magneto-hydrodynamics. JCP, **203**(1), 344–357, 10 February (2005)
23. Malika, R., Wagdi, H., A discontinuous Galerkin method with the HLLC solver for the Euler equations. IJNMF, **43**, (12), 1391–1405, Dic. (2003)
24. Mignone, A., & Bodo, G., An HLLC solver for relativistic flows II. Magneto-hydrodynamics. MNRAS, **368**, 1040–1054 (2006)

25. Miyoshia, T., & Kusanob, K., A multi-state HLL approximate Riemann solver for ideal magnetohydrodynamics. JCP, **208**(1), 315–344, 1 September, (2005)
26. Schwartzkopff, T., Munz, C.D., & Toro, E.F., ADER: High-order approach for linear hyperbolic systems in 2D. JSC, **17** (1–4), 231–240 (2002)
27. Smoller, J., Shock Waves and Reaction–Diffusion Equations, Springer-Verlag, Berlin (1994)
28. Sweby, P.K., High resolution schemes using flux limiters for hyperbolic conservation laws. SIAMJNA, **21**, 995–1011 (1984)
29. Takakura, Y., & Toro, E.F., Arbitrarily accurate non-oscillatory schemes for a non-linear conservation law. CFDJ, **11** (1), 7–18 (2002)
30. Titarev, V.A., & Toro, E.F., ADER: Arbitrary high. Order Godunov approach. JSC, **17** (1–4) 609–618, (2002)
31. Titarev, V.A., & Toro, E.F., ADER schemes for three-dimensional non-linear hyperbolic systems. JCP, **204**, 715–736 (2005)
32. Toro, E.F., Spruce, M., & Speares, W., Restoration of the contact surface in the HLL Riemann solver. Report CoA 9204, June 1992, Cranfield Institute of Technology, UK
33. Toro, E.F., Spruce, M., & Speares, W., Restoration of the contact surface in the HLL Riemann solver. JShWa, **4**, 25–34, 1994
34. Toro, E.F., & Chakraborty, A., Development of an approximate Riemann solver for the steady supersonic Euler equations. AJ, **98**, 325–339 (1994)
35. Toro, E.F., Riemann Solvers and Numerical Methods for Fluid Dynamics. A Practical Introduction. First Edition, Springer-Verlag, Berlin (1997)
36. Toro, E.F., Riemann Solvers and Numerical Nethods for Fluid Dynamics. A Practical Introduction. Second Edition, Springer-Verlag, Berlin (1999)
37. Toro, E.F., A Library of Source Codes for Teaching, Research and Applications, Numeritek Ltd (May 1999)
38. Toro, E.F., & Billett, S.J., Centred TVD schemes for hyperbolic conservations laws. IMAJNA, **20**, 47–79 (2000)
39. Toro, E.F., Millington, R.C., & Nejad, L.A.M., Towards very high-order godunov schemes. In: Godunov Methods: Theory and Applications, Edited Review, Toro, E.F. (Editor). Kluwer Academic/Plenum Publishers, pp. 905–938 (2001)
40. Toro, E.F., & Titarev, V.A., Solution of the generalised Riemann problem for advection-reaction equations. PRSLA, **458**, 271–281 (2002)
41. Toro, E.F., Multi-stage predictor-corrector fluxes for hyperbolic equations. Isaac Newton Institute for Mathematical Sciences Preprint Series, NI03037-NPA, University of Cambridge, UK, June 2003
42. Toro, E.F., & Titarev, V.A., ADER schemes for scalar hyperbolic conservation laws with source terms in three space dimensions. JCP, **202**, 196–215 (2005)
43. Toro, E.F., & Titarev, V.A., MUSTA fluxes for systems of conservation laws. JCP, **216**, 403–429 (2006)
44. Toro, E.F., A Riemann solver with evolved initial conditions. IJNMF, **52**, 433–453 (2006)
45. Toro, E.F., & Hidalgo, A., ADER finite volume schemes for nonlinear diffusion–reaction equations. Isaac Newton Istitute for Mathematical Sciences, Univesity of Cambridge, UK. Preprint NI-07011-NPA, 2007
46. Zachmanoglou, E.C., & Thoe, D.W., Introduction to Partial Differential Equations, Dover Publications. Inc. New York (1986)

Shock-Capturing Schemes in Computational MHD

A. Mignone[1,2] and G. Bodo[1]

[1] INAF Osservatorio Astronomico di Torino, 10025 Pino Torinese, Italy,
mignone@to.astro.it
[2] Dipartimento di Fisica Generale dell'Università, Via Pietro Giuria 1, I-10125
Torino, Italy, bodo@to.astro.it

Abstract. The purpose of the present review is to present and discuss some
introductory aspects relevant to computational compressible magnetohydrodynamics (MHD). The shock-capturing framework developed for the Euler equations of
gasdynamics is extended to MHD by illustrating differences and additional complexities introduced by the presence of magnetic fields. In particular, we focus our
attention on the characteristic structure of the equations by investigating the nature
of different MHD waves, the solution to the Riemann problem and last, but not
least, various computational strategies to control the divergence-free condition of
magnetic fields.

Keywords Magnetohydrodynamics (MHD) · Methods: numerical · Shock
waves · Waves

1 Introduction

Astrophysical plasmas can often be described by means of the ideal compressible magnetohydrodynamic (MHD) equations. A far from exhaustive list
includes jets, accretion disks, stellar or galactic atmospheres, and the interstellar medium. In many instances, one has to deal with flows with shocks
and discontinuities, and, in such situations, the numerical methods used in
the simulations are based on the shock-capturing framework developed for
the Euler equations of gasdynamics. The extension of such framework to
MHD has proven, however, to be nontrivial because of several properties
of the MHD system that makes it different from the Euler counterpart. A
first example of the problems encountered when moving from gasdynamic to
MHD is nonstrict hyperbolicity. This has been addressed by [5] and [40], and
following this and other advancements, several second-order upwind codes
were then constructed and tested mainly for the one-dimensional case, see
for example [2, 42, 50] and [10]. New problems have to be considered for the

multidimensional case, in particular MHD equations are supplemented by the condition of null divergence of the magnetic field that has to be preserved during the evolution. A failure of the numerical scheme to maintain this constraint, as shown by [4], leads to unphysical effects in the solution. Several different methods have then been proposed for dealing with this issue, see for example [3, 4, 12, 17, 19, 29, 36].

In this review, after a presentation of the equations in Sect. 2, we will discuss their characteristic structure and the nature of the waves with particular reference to the peculiarities of the MHD system in Sect. 3. The solution to the Riemann problem is one of the main building blocks of shock capturing methods, and our discussion will be focused on approximate solvers, in particular, of the HLL class. In the last section, we will deal with the other big issue of computational MHD and will present several different computational strategies for approaching the divergence free constraint.

2 The MHD Equations

The macroscopic dynamics of a plasma can be described, in many instances, by the MHD equations. Describing the plasma, as a single fluid, in terms of density ρ, velocity $\boldsymbol{v}$, thermal pressure p_g, and magnetic field $\mathbf{B}$, the MHD equations take the form

$$\frac{\partial \rho}{\partial t} = -\nabla \cdot (\rho \mathbf{v}), \tag{1}$$

$$\rho \frac{\partial \mathbf{v}}{\partial t} = -\rho \mathbf{v} \cdot \nabla \mathbf{v} - \nabla p_g + \mathbf{J} \times \mathbf{B}, \tag{2}$$

$$\frac{\partial p_g}{\partial t} = -\nabla \cdot (p_g \boldsymbol{v}) - (\Gamma - 1) \left(p_g \nabla \cdot \boldsymbol{v} - \eta \mathbf{J}^2 \right), \tag{3}$$

$$\frac{\partial \mathbf{B}}{\partial t} = -\nabla \times \boldsymbol{\Omega}, \tag{4}$$

$$\mathbf{J} = \nabla \times \mathbf{B}, \tag{5}$$

$$\boldsymbol{\Omega} = -\mathbf{v} \times \mathbf{B} + \eta \mathbf{J}, \tag{6}$$

where $\boldsymbol{\Omega}$ denotes the electric field. The units of $\mathbf{B}$ are chosen such that the magnetic permeability of vacuum becomes equal to unity, $\mu_0 = 1$. In Eq. (5), expressing the current density $\mathbf{J}$ in terms of the magnetic field, we neglected the displacement current $\partial \boldsymbol{\Omega}/\partial t$. This is justified if we are far from the relativistic regime, i.e., $v \ll c$. Eq. (6) is the Ohm's law and η is the resistivity. With respect to the Euler system of gasdynamics, we have, in Eq. (2), the additional term $\mathbf{J} \times \mathbf{B}$ that represents the Lorentz force and in Eq. (3) the Joule heating term $\eta \mathbf{J}^2$. Moreover, we have the additional induction equation (4) for the evolution of the magnetic field. The condition $\nabla \cdot \mathbf{B} = 0$

represents a constraint that has to be satisfied by the magnetic field at all times, and Eq. (4) ensures that if it is fulfilled at $t = 0$ it will be so at any time. In system (1, 2, 3, 4, 5, 6) and in the rest of the Section, we keep the resistive terms for completeness although we will not discuss their numerical treatment.

Shock capturing methods are based on the conservation form that, for the above system, can be written as:

$$\frac{\partial \rho}{\partial t} = -\nabla \cdot (\rho \mathbf{v}) , \tag{7}$$

$$\frac{\partial \rho \mathbf{v}}{\partial t} = -\nabla \cdot \left(\rho \boldsymbol{vv} + \mathbf{I} p_g + \mathbf{I} \frac{\mathbf{B}^2}{2} - \mathbf{BB} \right) , \tag{8}$$

$$\frac{\partial E}{\partial t} = -\nabla \cdot \left(\mathbf{v} \left(E + p_g \right) + \mathbf{v} \cdot \left(\mathbf{I} \frac{\mathbf{B}^2}{2} - \mathbf{BB} \right) - \mathbf{B} \times \eta \mathbf{J} \right) , \tag{9}$$

$$\frac{\partial \mathbf{B}}{\partial t} = \nabla \times (\mathbf{v} \times \mathbf{B} - \eta \mathbf{J}) , \tag{10}$$

where $\mathbf{I}$ is the unit dyadic and the total energy density E is defined as

$$E = \frac{p_g}{\Gamma - 1} + \frac{\rho v^2}{2} + \frac{\mathbf{B}^2}{2} . \tag{11}$$

The above conservation equations can be written in the compact form

$$\frac{\partial \mathbf{U}}{\partial t} + \nabla \cdot \mathbf{F} = \mathbf{0} , \tag{12}$$

where the vector $\mathbf{U}$ of conserved quantities and the vector $\mathbf{F}$ of fluxes, neglecting resistivity, are given, respectively, by

$$\mathbf{U} = \begin{pmatrix} \rho \\ \rho \boldsymbol{v} \\ E \\ \mathbf{B} \end{pmatrix} , \quad \mathbf{F} = \begin{bmatrix} \rho \boldsymbol{v} \\ \rho \boldsymbol{vv} + \mathbf{I} p - \mathbf{BB} \\ (E + p) \boldsymbol{v} - (\boldsymbol{v} \cdot \mathbf{B}) \mathbf{B} \\ \boldsymbol{v} \mathbf{B} - \mathbf{B} \boldsymbol{v} \end{bmatrix} , \tag{13}$$

where $p = p_g + \mathbf{B}^2/2$ denotes the total pressure.

3 The Riemann Problem in MHD

The solution to the Riemann problem in magnetohydrodynamics is paved by several additional complications when compared to the underlying hyperbolic system of the Euler equation of gasdynamics. The increased complexity cannot be ascribed just to the increased number of waves due to a larger number of equations, but also to the fact that the system is non strictly hyperbolic with non convex flux function, and the characteristic fields are no longer either genuinely non linear or linearly degenerated. In this respect, non regular waves like compound waves and overcompressive intermediate shocks may be formed in MHD.

As usual, one starts with the problem definition, i.e., a discontinuity separating a pair of arbitrary constant left and right states,

$$\boldsymbol{U}(x, t = 0) = \begin{cases} \boldsymbol{U}_{\mathrm{L}} & \text{for} \quad x < 0, \\ \boldsymbol{U}_{\mathrm{R}} & \text{for} \quad x > 0. \end{cases} \tag{14}$$

As with hydrodynamics Riemann solvers, the initial jump will decay into a set of uniform states separated by left- and right-facing shock and rarefaction waves. In general, at $t > 0$, the full structure comprises a total of eight states (including the original ones) separated by seven waves, see Fig. 1. With the exception of the entropy mode associated with a contact discontinuity moving at the speed of the fluid, the other six waves are related to fast, Alfvén, and slow characteristics and can be either shocks or rarefactions. In addition to this, two families of waves may occasionally have the same speeds or develop compound wave structures where both a shock and a rarefaction propagate

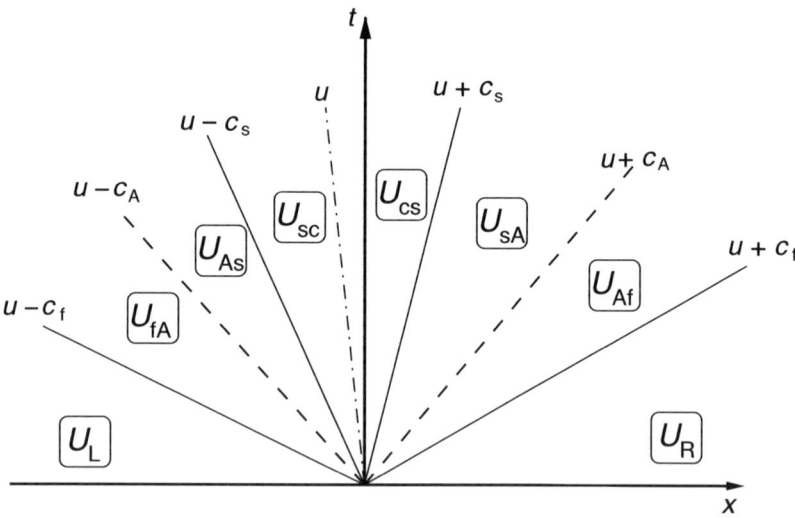

Fig. 1. General structure of the Riemann fan generated by two initial constant state $\boldsymbol{U}_{\mathrm{L}}$ and $\boldsymbol{U}_{\mathrm{R}}$. The pattern comprises seven waves corresponding to a pair of fast ($u \pm c_{\mathrm{f}}$), Alfvén ($u \pm c_{\mathrm{A}}$), slow ($u \pm c_{\mathrm{s}}$) modes separated by a contact discontinuity in the middle (u). The waves bound six new constant states, from left to right, $\boldsymbol{U}_{\mathrm{fA}}$, $\boldsymbol{U}_{\mathrm{As}}$, $\boldsymbol{U}_{\mathrm{sc}}$, $\boldsymbol{U}_{\mathrm{cs}}$, $\boldsymbol{U}_{\mathrm{sA}}$, $\boldsymbol{U}_{\mathrm{Af}}$

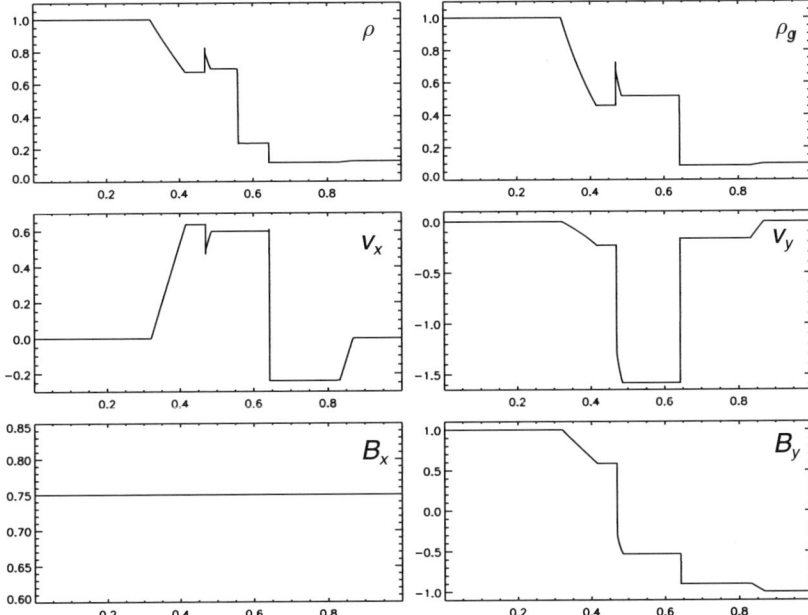

Fig. 2. The Brio-Wu shock tube problem. From left to right, the solution involves five waves: a *fast rarefaction wave*, a *slow compound wave*, a *contact wave*, a slow shock, and a fast rarefaction. Here $\Gamma = 2$. Density and pressure are shown in the *top* panels, velocities in the *middle* ones, and magnetic fields in the *bottom* panels. Results have been obtained using an adaptive grid with an equivalent resolution of 12, 800 grid zones

adjacent to one another. An example of such situation is encountered in the Brio and Wu [5] shock tube problem, with initial left and right data given by $(\rho, p_g)_L = (1, 1)$, $\mathbf{B}_L = (3/4, 1, 0)$, $\boldsymbol{v}_L = \boldsymbol{v}_R = \mathbf{0}$, $(\rho, p_g)_R = (1/8, 1/10)$ and $\boldsymbol{B}_R = (3/4, -1, 0)$. The results computed with the PLUTO code [34] are shown in Fig. 2: a compound structure consisting of a slow compressive shock and a rarefaction is clearly visible at $x \approx 0.46$. This peculiar behavior is a direct consequence of the fact that the MHD system of equations are neither strictly hyperbolic nor strictly convex. For these reasons, unlike the Euler equations of gasdynamics, a closed form solution to the Riemann problem in MHD cannot be found.

Instead, accurate solvers may be constructed in the regular case by proceeding as for the hydrodynamical problem, [7, 8, 9, 18, 42, 50]. This demands the simultaneous solution of the Rankine-Hugoniot jump conditions across each wave by a self-consistent procedure that resolves fast, slow, and Alfvén waves to the left and right of the contact (tangential in the degenerate case) discontinuity, always located at the center of the structure.

3.1 Discontinuities

The jump conditions can be derived directly from the original conservation
law by integrating across a discontinuous surface. They are better expressed
in Lagrangian mass coordinates,

$$
\begin{aligned}
W \left[\frac{1}{\rho} \right] &= -[u] \,, \\
W [u] &= [p] \,, \\
W [\boldsymbol{v}_t] &= -B_x [\boldsymbol{B}_t] \,, \\
W \left[\frac{\boldsymbol{B}_t}{\rho} \right] &= -B_x [\boldsymbol{v}_t] \,, \\
W \left[\frac{E}{\rho} \right] &= [up] - B_x [\boldsymbol{B} \cdot \boldsymbol{v}] \,,
\end{aligned}
\tag{15}
$$

where W is the mass flux entering a discontinuity surface and $[\cdot]$ is the difference between the two states on each side of the front.

In the previous equations, $p = p_g + \boldsymbol{B}^2/2$ denotes the total pressure, u and
B_x are the projections of the velocity and magnetic field vectors on the discontinuity normal, and the subscript t refers to the tangential components. Note
also that the normal component of magnetic field, B_x, is constant throughout
the Riemann fan and should be regarded as a parameter.

Rotational waves are linear waves characterized by

$$
[u] = [p_g] = \left[\boldsymbol{B}_t^2 \right] = [\rho] = 0 \,.
\tag{16}
$$

Thus thermodynamical quantities such as density and pressure remain un-
altered across the discontinuity and only tangential components can change.
Indeed, vector fields experience a rotation without changing their magnitude,
meaning that the total pressure is also continuous, i.e., $[p] = 0$. For this reason, a rotational discontinuity cannot be formed by the steepening of a smooth
disturbance. Rotational waves propagate at the Alfvén speed relative to the
fluid, $u \pm B_x/\sqrt{\rho}$.

Shock waves are physically admissible if the entropy is increased through
the front. Through a shock wave, all hydrodynamics variables including density, pressure, velocity, and magnetic fields are subjected to a jump. From the
third and fourth equations in (15), one sees that the magnetic field on both
sides of the shock lies on the same plane and no rotation takes place. Thus
only the magnitude of the field can change.

A fast shock is characterized by an increased magnitude of the transverse magnetic field when passing from the pre shock to the post shock
state [7, 25]. This has the consequence to bend the field lines away from
the shock normal. The downstream magnetic field, however, is not a monotonic function of the Mach number for sufficiently large values of B_x. This

means that the transverse component of $\boldsymbol{B}$ in the downstream state does not determine the post shock values uniquely [18, 25].

Through a slow shock, on the contrary, the magnitude of the transverse component of magnetic field decreases from the upstream to the downstream state. In this case, the magnetic field bends towards the shock normal. Slow shocks are more peculiar than fast shocks [47], since all the familiar quantities (i.e., pressure, density, and transverse field) do not behave in a monotonic way. Furthermore, the range of Mach number values for which a slow shock can exist is finite.

Shocks propagating in the direction of the magnetic field are called parallel shocks. In this case, $\boldsymbol{B}_t$ vanishes on both sides of the discontinuity (although $B_x \neq 0$) and the shock becomes purely hydrodynamical. In the case of a fast shock, however, another solution exists where the tangential magnetic field vanishes ahead of the front but $\boldsymbol{B}_t \neq 0$ in the downstream region. Such a particular configuration is called a *switch-on* shock, since the magnetic field is "turned-on" behind the shock. In the frame of the front, the downstream fluid moves at the local Alfvén velocity. It is worth mentioning that switch-on shocks only exist in a small range of upstream parameters [26], namely $B_x^2 > \Gamma p$ and

$$1 < M_A < \sqrt{\frac{\Gamma(1 - \beta) + 1}{\Gamma - 1}}, \tag{17}$$

where $M_A = v/c_A$ is the Alfvénic Mach number and $\beta = 2p/B_x^2$ is the plasma β. A reverse situation is encountered in the case of a slow shock; in this case $\boldsymbol{B}_t$ is zero in the downstream region, while it does not vanish in the upstream. This configuration corresponds to a *switch-off* shock, since the post shock magnetic field is switched off. Such fronts propagate at the Alfvén speed of the upstream medium.

When $B_x = 0$, we have a perpendicular (normal) shock. From the jump conditions (15), one immediately sees that the tangential velocity does not change through the discontinuity and the magnetic field is compressed by the same ratio as for the density, without changing its direction.

Figure 3 shows the wave pattern emerging from an initial jump [2] separating a left state with $(\rho, p)_L = (1.08, 0.95)$, $\boldsymbol{v}_L = (1.2, 0.01, 0.5)$, $\boldsymbol{B}_L = (2, 3.6, 2)/\sqrt{4\pi}$ from a right state with $(\rho, p)_R = (1, 1)$, $\boldsymbol{v}_R = \boldsymbol{0}$, $\boldsymbol{B}_R = (2, 4, 2)/\sqrt{4\pi}$. As it can be seen, the solution involves discontinuities only; two opposite moving fast shocks followed by two rotational waves and slow shocks. The contact discontinuity is located at the center of the structure. Computations have been carried with $\Gamma = 5/3$ using the PLUTO code together with the Roe Riemann solver described below.

3.2 Rarefaction Waves

Fast and slow modes allow gas expansion by rarefaction waves. Across them, flow variables experience a smooth transition and the admissible states may be

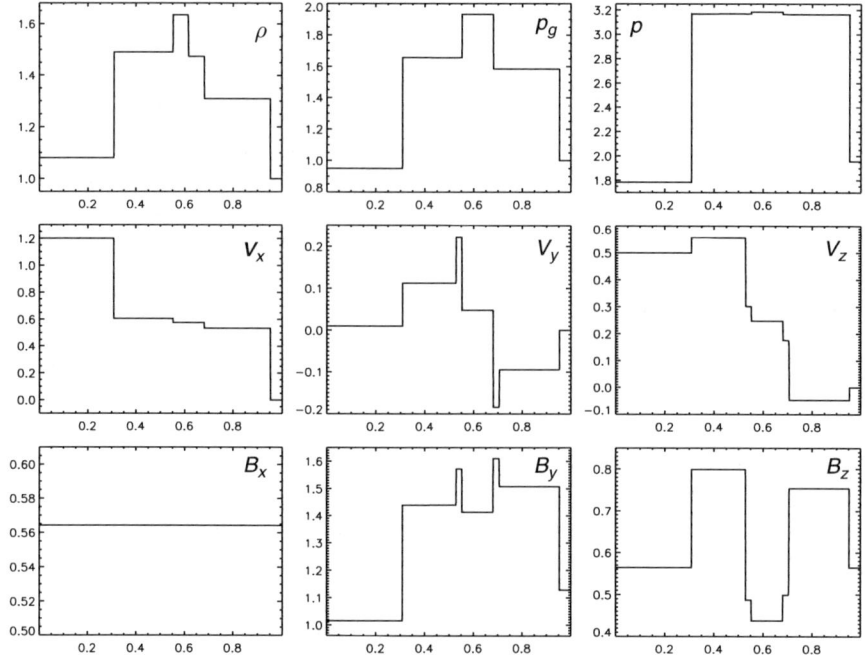

Fig. 3. A shock tube problem. Density, thermal, and total pressures are plotted in the *top* panels. Velocity and magnetic field components are shown in the *middle* and *bottom* panels, respectively. The outcoming wave structure involves two outermost fast shocks, enclosing two *rotational waves* containing two slow shocks separated by a *contact wave*

found from the integral curves which follow the eigenvectors of the hyperbolic system. Given the decomposition of the Jacobian matrix $\partial \boldsymbol{F}(\boldsymbol{U})/\partial \boldsymbol{U}$ in terms of right and left ortho-normal eigenvectors $\boldsymbol{R}_k$, $\boldsymbol{L}_k$ ($\boldsymbol{R}_k \cdot \boldsymbol{L}_m = \delta_{km}$) associated with eigenvalues λ_k, one can proceed as for the hydrodynamical case, that is, by replacing the set of jump conditions (15) with a set of ordinary differential equations:

$$\frac{d\boldsymbol{U}}{d\sigma} = \boldsymbol{R}_k \,, \tag{18}$$

where σ is a parameter along the curve. Equation (18) implies conservation of all Riemann invariants

$$dw_m \equiv \boldsymbol{L}_m \cdot d\boldsymbol{U} = 0 \qquad \text{for} \qquad m \neq k \,, \tag{19}$$

not associated with the wave involved. It can be shown [47] that, in contrast to a shock wave, the transverse components of the magnetic field always decrease through a fast rarefaction and increase through a slow one. As for the hydrodynamical case, the entropy is constant throughout the Riemann fan.

3.3 Approximate Riemann Solvers

In what follows, we will restrict our attention to the one-dimensional set of MHD equations written in conservation form,

$$\frac{\partial \boldsymbol{U}}{\partial t} + \frac{\partial \boldsymbol{F}}{\partial x} = 0\,, \tag{20}$$

with conservative variables and fluxes given, respectively, by

$$\boldsymbol{U} = \begin{pmatrix} \rho \\ \rho u \\ \rho \boldsymbol{v}_t \\ B_x \\ \boldsymbol{B}_t \\ E \end{pmatrix}, \quad \boldsymbol{F} = \begin{pmatrix} \rho u \\ \rho u^2 + p - B_x^2 \\ \rho \boldsymbol{v}_t u - \boldsymbol{B}_t B_x \\ 0 \\ \boldsymbol{B}_t u - \boldsymbol{v}_t B_x \\ (E + p)u - (\boldsymbol{v} \cdot \boldsymbol{B})B_x \end{pmatrix}. \tag{21}$$

The discrete version of Eq. (20) takes the difference form

$$\boldsymbol{U}_i^{n+1} = \boldsymbol{U}_i^n - \frac{\Delta t}{\Delta x}\left(\hat{\boldsymbol{f}}_{i+\frac{1}{2}} - \hat{\boldsymbol{f}}_{i-\frac{1}{2}}\right). \tag{22}$$

The numerical flux functions $\hat{\boldsymbol{f}}$ are computed at each cell interface $i + \frac{1}{2}$ by solving a Riemann problem between suitable left and right input states. For a first-order scheme, these states are provided by $\boldsymbol{U}_i^n$ and $\boldsymbol{U}_{i+1}^n$, respectively.

Nonlinear methods are based on the simultaneous solution of the Rankine Hugoniot jump conditions (15) or the smooth relations (18) across all waves in the system. The strategy proceeds by linking each constant state inside the Riemann fan with the next adjacent one, by exploiting the properties of the wave separating them. Approximate nonlinear solutions have been presented by [7, 8, 9], which treat rarefaction waves as discontinuities ("rarefaction shocks"). This simplification is sufficiently accurate in the limit of weak rarefactions and/or small time steps, such as the ones typically used in explicit methods. Exact Riemann solvers that correctly treat rarefaction waves have been proposed by [18, 42].

In either case (exact or approximate), nonlinear methods are rather involved and computationally intensive. This has motivated the probe of simplified, more efficient strategies of solution based on different levels of approximation. Indeed, the most popular methods nowadays adopted for the solution of the Riemann problem in computational MHD lean on approximate solvers. In what follows, we present a brief overview of the most popular approaches adopted.

Lax-Friedrichs and Rusanov Solvers

Perhaps the simplest approximation comes from the Lax-Friedrichs or Rusanov fluxes. The original Lax-Friedrichs method results from the attempt of stabilizing the unstable FTCS (forward in time, central in space) discretization. Specifically, one has

$$\hat{f} = \frac{1}{2} \left[F_L + F_R - c \left(U_R - U_L \right) \right], \tag{23}$$

where $F_L \equiv F(U_L)$ and $F_R \equiv F(U_R)$. Choosing $c = \Delta x / \Delta t$ yields the original Lax-Friedrichs scheme [27]. A considerably less diffusive variant is given by the Rusanov flux [41], where c is taken to be the maximum signal velocity $|\lambda_{\max}|$. For the MHD equations, the obvious choice is

$$\lambda_{\max} = |u| + c_f, \tag{24}$$

where both u and c_f may be evaluated using the average state $U_{RL} = (U_L + U_R)/2$.

The Rusanov flux is computationally inexpensive and straightforward to implement, since it does not require any characteristic information. Numerical experience shows that the scheme is quite robust and well-behaved, although rather diffusive, since all the knowledge about the wave structure inside the Riemann problem is avoided.

The Scheme of Harten–Lax–van Leer (HLL)

The HLL method, originally devised by Harten, Lax, and van Leer [22] for classical gasdynamics, has gained increasing popularity among researchers in the last decades.

The HLL scheme is formulated in terms of an integral average across the Riemann fan provided the leftmost (λ_L) and rightmost (λ_R) signal speeds can be estimated. This leads to an approximation of the Riemann fan structure where all the intermediate wave patterns are averaged into a single constant state bounded by two outermost waves, see Fig. 4. In other words, the solution to the Riemann problem on the $x/t = 0$ axis consists of three possible constant states:

$$U(0, t) = \begin{cases} U_L & \text{if} \quad \lambda_L \geq 0, \\ U^{\text{hll}} & \text{if} \quad \lambda_L \leq 0 \leq \lambda_R, \\ U_R & \text{if} \quad \lambda_R \leq 0. \end{cases} \tag{25}$$

The single state U^{hll} is constructed from an a priori estimate of the fastest and slowest signal velocities λ_L and λ_R:

$$U^{\text{hll}} = \frac{\lambda_R U_R - \lambda_L U_L + F_L - F_R}{\lambda_R - \lambda_L}, \tag{26}$$

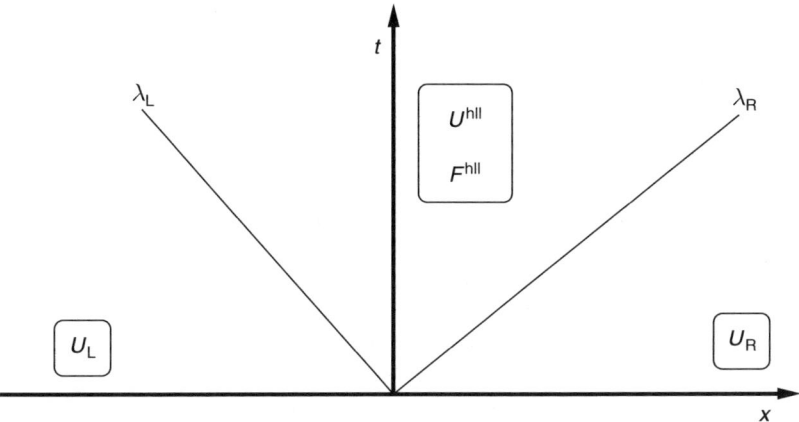

Fig. 4. Approximate structure of the Riemann fan used in the HLL solver: the whole fan has been lumped into a single state U^{hll} and flux F^{hll}

where $F_{\mathrm{L}} = F(U_{\mathrm{L}})$, $F_{\mathrm{R}} = F(U_{\mathrm{R}})$. Notice that Eq. (26) represents the integral average of the solution of the Riemann problem over the wave fan [46]. Demanding consistency with the jump condition across either of the bounding waves, the corresponding interface numerical flux can be derived as:

$$\hat{f} = \begin{cases} F_{\mathrm{L}} & \text{if} \quad \lambda_{\mathrm{L}} \geq 0, \\ F^{\mathrm{hll}} & \text{if} \quad \lambda_{\mathrm{L}} \leq 0 \leq \lambda_{\mathrm{R}}, \\ F_{\mathrm{R}} & \text{if} \quad \lambda_{\mathrm{R}} \leq 0, \end{cases} \tag{27}$$

where

$$F^{\mathrm{hll}} = \frac{\lambda_{\mathrm{R}} F_{\mathrm{L}} - \lambda_{\mathrm{L}} F_{\mathrm{R}} + \lambda_{\mathrm{R}} \lambda_{\mathrm{L}} (U_{\mathrm{R}} - U_{\mathrm{L}})}{\lambda_{\mathrm{R}} - \lambda_{\mathrm{L}}}. \tag{28}$$

Thus, given an estimate for the fastest and slowest signal speeds λ_{R} and λ_{L}, an approximate solution to the Riemann problem can be constructed and the intercell numerical flux is computed according to (27). Note that, in the supersonic case ($\lambda_{\mathrm{L}} > 0$ or $\lambda_{\mathrm{R}} < 0$), the HLL approximation gives the exact solution by selecting the correct upwind flux.

The algorithm is complete once λ_{L} and λ_{R} have been specified. Several estimates have been proposed, see for example [15, 16, 45, 46]. A popular choice [15], for example, is to compute the signal velocities using the data available in the left and right states, and then define

$$\lambda_{\mathrm{L}} = \min \left[\lambda_-(U_{\mathrm{L}}), \lambda_-(U_{\mathrm{R}}) \right], \quad \lambda_{\mathrm{R}} = \max \left[\lambda_+(U_{\mathrm{L}}), \lambda_+(U_{\mathrm{R}}) \right]. \tag{29}$$

The HLL approach does not require a full characteristic decomposition of the equations and, for this reason, it is straightforward to implement in any functioning MHD code. Besides its computational efficiency and ease of implementation, the HLL scheme has the attractive feature of being positively conservative in the sense that preserve initially positive densities, energies, and pressures. Despite its reliability, however, it lacks the ability to resolve intermediate structures such as Alfvén, slow, and contact modes. This results in a more diffusive behavior than other more sophisticated algorithms.

The HLLC Approximate Riemann Solver

The HLLC scheme [1, 45] improves over HLL by replacing the single averaged state defined by (26) with two approximate states, U_L^* and U_R^*. These two states are separated by a middle wave which is assumed to have constant speed λ^*,

$$
U(0,t) = \begin{cases} U_L & \text{if} & \lambda_L \geq 0, \\ U_L^* & \text{if} & \lambda_L \leq 0 \leq \lambda^*, \\ U_R^* & \text{if} & \lambda^* \leq 0 \leq \lambda_R, \\ U_R & \text{if} & \lambda_R \leq 0, \end{cases} \tag{30}
$$

and the corresponding inter cell numerical fluxes become

$$
\hat{f} = \begin{cases} F_L & \text{if} & \lambda_L \geq 0, \\ F_L^* & \text{if} & \lambda_L \leq 0 \leq \lambda^*, \\ F_R^* & \text{if} & \lambda^* \leq 0 \leq \lambda_R, \\ F_R & \text{if} & \lambda_R \leq 0. \end{cases} \tag{31}
$$

This configuration is schematically depicted in Fig. 5.

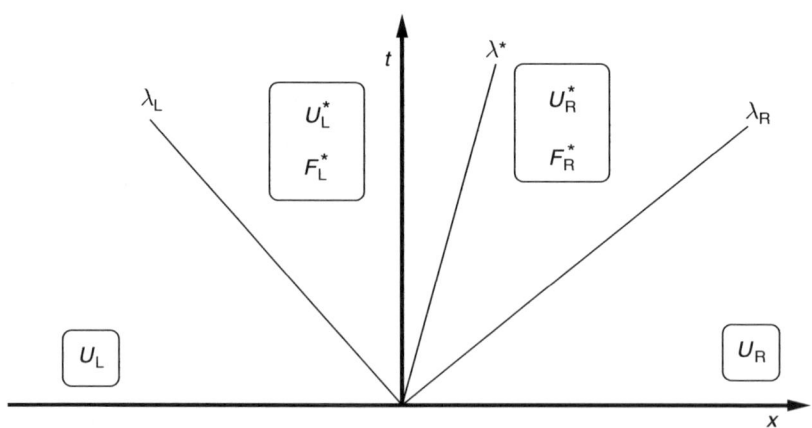

Fig. 5. Approximate structure to the Riemann fan used by the HLLC solver: the whole fan has been reduced to two single states, U_L^* and U_R^*, separated by a *middle wave* λ^*

Note that, even if $\boldsymbol{F}_\alpha \equiv \boldsymbol{F}(\boldsymbol{U}_\alpha)$ for $\alpha = \{L, R\}$, one cannot take $\boldsymbol{F}_\alpha^* = \boldsymbol{F}(\boldsymbol{U}_\alpha^*)$. The intermediate fluxes $\boldsymbol{F}_L^*$ and $\boldsymbol{F}_R^*$, in fact, should be computed consistently from the Rankine-Hugoniot jump conditions across each wave:

$$\lambda_\alpha \left(\boldsymbol{U}_\alpha^* - \boldsymbol{U}_\alpha\right) = \boldsymbol{F}_\alpha^* - \boldsymbol{F}_\alpha \,, \tag{32}$$

where $\alpha = L$ or $\alpha = R$ for the left or right state, respectively. In this respect, $\boldsymbol{F}_L^*$ and $\boldsymbol{F}_R^*$ should be regarded as independent unknowns in the problem. Additionally, if the middle wave is taken to be a contact discontinuity, states and fluxes should also satisfy the jump conditions across it,

$$\lambda^* \left(\boldsymbol{U}_L^* - \boldsymbol{U}_R^*\right) = \boldsymbol{F}_L^* - \boldsymbol{F}_R^* \,, \tag{33}$$

by consistently demanding continuity of magnetic field, velocity, and total pressure across λ^*. Equation (32) may be replaced by two alternative sets derived by direct summation of the jump relations across all waves [31, 32, 33].

The first set yields the traditional consistency conditions [46] in terms of the state vectors $\boldsymbol{U}_L^*$ and $\boldsymbol{U}_R^*$,

$$\frac{(\lambda^* - \lambda_L)\boldsymbol{U}_L^* + (\lambda_R - \lambda^*)\boldsymbol{U}_R^*}{\lambda_R - \lambda_L} = \boldsymbol{U}^{\mathrm{hll}} \,, \tag{34}$$

where $\boldsymbol{U}^{\mathrm{hll}}$ is given by Eq. (26). The second set, obtained after dividing each jump condition by the corresponding wave speed and adding the resulting expressions, yields a similar relation for the fluxes $\boldsymbol{F}_L^*$ and $\boldsymbol{F}_R^*$:

$$\frac{\boldsymbol{F}_L^* \lambda_R(\lambda^* - \lambda_L) + \boldsymbol{F}_R^* \lambda_L(\lambda_R - \lambda^*)}{\lambda_R - \lambda_L} = \lambda^* \boldsymbol{F}^{\mathrm{hll}} \,, \tag{35}$$

where $\boldsymbol{F}^{\mathrm{hll}}$ is given by Eq. (28).

The problem is well posed if the number of unknowns exactly matches the number of available equations. Considering the normal component of the magnetic field, B_x, as a constant parameter, one has at disposal a total of 21 equations: 14 for the outer waves $(7+7)$ and 7 from the jumps across the middle contact wave, Eq. (33). Since λ^* is unknown (it is not determined a priori), it follows that states and fluxes in the *star* region should be written in terms of 20 unknown variables, 10 per state. There is, however, some degree of freedom in choosing this representation.

Both Gurski [21] and Li [28] proposed to express the unknowns and fluxes in the *star* regions as

$$\boldsymbol{U}_\alpha^* = \begin{pmatrix} \rho_\alpha^* \\ \rho_\alpha^* \lambda^* \\ \rho_\alpha^* \boldsymbol{v}_{t_\alpha}^* \\ \boldsymbol{B}_{t_\alpha}^* \\ E_\alpha^* \end{pmatrix}, \quad \boldsymbol{F}_\alpha^* = \begin{pmatrix} \rho_\alpha^* \lambda^* \\ \rho_\alpha^* (\lambda^*)^2 + p^* - B_x^2 \\ \rho_\alpha^* \boldsymbol{v}_{t_\alpha}^* \lambda^* - B_x \boldsymbol{B}_{t_\alpha}^* \\ \boldsymbol{B}_{t_\alpha}^* \lambda^* - \boldsymbol{v}_{t_\alpha}^* B_x \\ (E_\alpha^* + p_\alpha^*)\lambda^* - B_x \left(\boldsymbol{B} \cdot \boldsymbol{v}\right)_\alpha^* \end{pmatrix}, \tag{36}$$

where $\alpha = L$ or $\alpha = R$ for the left or right state, respectively. In Eq. (36) the fluid normal velocity is assumed to be continuous across the contact mode and equal to the speed of the discontinuity itself, that is, $\lambda^* = u_L^* = u_R^*$. By exploiting the continuity of tangential magnetic field and total pressure across the contact (entropy) mode, one can easily find, from Eqs. (34) and (35), the *unique* choices

$$\lambda^* = \frac{m_x^{\mathrm{hll}}}{\rho^{\mathrm{hll}}}, \quad \boldsymbol{B}_{t_L}^* = \boldsymbol{B}_{t_R}^* = \boldsymbol{B}_t^{\mathrm{hll}}, \quad p_L^* = p_R^* = F_{[m_x]}^{\mathrm{hll}} + B_x^2 - F_{[\rho]}^{\mathrm{hll}}\lambda^*, \quad (37)$$

where m_x^{hll} and $\boldsymbol{B}_{t_\alpha}^{\mathrm{hll}}$ are, respectively, the normal momentum and transverse magnetic field components given by the HLL-averaged state $\boldsymbol{U}^{\mathrm{hll}}$, see Eq. (26).

In [28], the expressions for density and transverse components of momentum are derived from the jump conditions across the outermost waves, Eq. (32):

$$\rho_\alpha^* = \rho_\alpha \frac{\lambda_\alpha - u_\alpha}{\lambda_\alpha - \lambda^*}, \quad (\rho_\alpha \boldsymbol{v}_{t_\alpha})^* = \rho_\alpha^* \boldsymbol{v}_{t_\alpha} - B_x \frac{\boldsymbol{B}_{t_\alpha}^* - \boldsymbol{B}_{t_\alpha}}{\lambda_\alpha - \lambda^*}, \quad (38)$$

and similarly for the energy:

$$E_\alpha^* = \frac{E_\alpha(\lambda_\alpha - u_\alpha) + (p^*\lambda^* - p_\alpha u_\alpha) - B_x\left[(\boldsymbol{B} \cdot \boldsymbol{v})_\alpha^* - \boldsymbol{B}_\alpha \cdot \boldsymbol{v}_\alpha\right]}{\lambda_\alpha - \lambda^*}, \quad (39)$$

where, from the consistency condition, one must have $(\boldsymbol{B} \cdot \boldsymbol{v})_L^* = (\boldsymbol{B} \cdot \boldsymbol{v})_R^*$. Since $\boldsymbol{B}_{tL}^* = \boldsymbol{B}_{tR}^*$ is continuous across the middle wave, a possibility is to set $\boldsymbol{v}^* = \boldsymbol{m}^{\mathrm{hll}}/\rho^{\mathrm{hll}}$ as done in [28]. However, this choice is somewhat inconsistent with the expression of $\boldsymbol{v}_{tL}^*$ and $\boldsymbol{v}_{tR}^*$ recovered from the second equation in (38). Indeed, Li's formulation introduces only nine unknowns per state: $\rho^*, \lambda^*, \boldsymbol{v}_t^*, \boldsymbol{B}_t^*, E^*, p^*$, and $(\boldsymbol{v} \cdot \boldsymbol{B})^*$. As such, it has been derived by using 18 equations only, namely the 14 jumps across the outer waves, Eq. (32), together with the imposed continuity of u^*, $\boldsymbol{B}_t^*$, and p^*. Consequently, it fails to satisfy some of the jump relations across the middle wave as it can be verified from the components of the induction equation.

Similar inconsistencies can also be found in [21], which defines both normal and tangential velocities as ratios between HLL-averaged momentum and density, i.e., $\boldsymbol{v}_L^* \equiv \boldsymbol{v}_R^* = \boldsymbol{m}^{\mathrm{hll}}/\rho^{\mathrm{hll}}$. In a second derivation, in the attempt to gain further benefits from the introduction of the middle wave (such as capturing isolated slow or Alfvén waves), Gurski [21] relaxes the assumption of continuity in the transverse component of $\boldsymbol{v}$ and $\boldsymbol{B}$. The resulting $\boldsymbol{U}_L^*$ and $\boldsymbol{U}_R^*$ derived in this way do not satisfy the jump conditions. This freedom comes at the extra cost of introducing a dissipation terms to control unwanted spurious numerical oscillations.

A consistent formulation may be derived by following [32], who extended the HLLC formalism to the equations of relativistic MHD. Since one must have 10 unknowns per state, the following expressions can be introduced

$$
\boldsymbol{U}_\alpha^* = \begin{pmatrix} \rho_\alpha^* \\ \rho_\alpha^* \lambda^* \\ \boldsymbol{m}_{t_\alpha}^* \\ \boldsymbol{B}_{t_\alpha}^* \\ E_\alpha^* \end{pmatrix}, \quad
\boldsymbol{F}_\alpha^* = \begin{pmatrix} \rho_\alpha^* \lambda^* \\ \rho_\alpha^* (\lambda^*)^2 + p^* - B_x^2 \\ \boldsymbol{m}_{t_\alpha}^* \lambda^* - B_x \boldsymbol{B}_{t_\alpha}^* \\ \boldsymbol{B}_{t_\alpha}^* \lambda^* - \boldsymbol{v}_{t_\alpha}^* B_x \\ (E_\alpha^* + p_\alpha^*)\lambda^* - B_x \left(\boldsymbol{B} \cdot \boldsymbol{v} \right)_\alpha^* \end{pmatrix}, \quad (40)
$$

where $\boldsymbol{m}_{t_\alpha}^*$ is the transverse momentum. One must realize that, for the sake of consistency, one has to assume $\rho_\alpha^* \boldsymbol{v}_{t_\alpha}^* \neq \boldsymbol{m}_{t_\alpha}^*$ in the previous equations, although one still has $m_{x_\alpha}^* = \rho_\alpha^* \lambda^*$. This comes from the fact that the average momentum is not simply equal to the average density times the average velocity. Note that the proposed formulation is not necessarily unique; for example, there is some degree of freedom in writing term $(\boldsymbol{v} \cdot \boldsymbol{B})^*$ in the energy flux as $\boldsymbol{v}^* \cdot \boldsymbol{B}^*$ or $\boldsymbol{m}^* \cdot \boldsymbol{B}^*/\rho^*$. Nevertheless, the advantage offered by the previous expressions is the introduction of 10 unknowns per state, complemented by 14 equations across the outer fast shocks and six additional constraints imposed through the entropy mode:

$$
[p^*] = [\lambda^*] = [\boldsymbol{B}_{t_\alpha}^*] = [\boldsymbol{v}_{t_\alpha}^*] = 0. \quad (41)
$$

As for Li's solver, this leads to the unique choices given by Eq. (37). However, although the first equation for ρ_α^* in (38) still holds, the second one should be replaced by:

$$
\boldsymbol{m}_{t_\alpha}^* = \boldsymbol{m}_{t_\alpha} \frac{\lambda_\alpha - u_\alpha}{\lambda_\alpha - \lambda^*} + B_x \frac{\boldsymbol{B}_{t_\alpha} - \boldsymbol{B}_{t_\alpha}^*}{\lambda_\alpha - \lambda^*}. \quad (42)
$$

Having defined the magnetic field, the velocities can be derived directly from Eq. (35),

$$
\boldsymbol{v}_{t_\alpha}^* = \frac{\boldsymbol{B}_{t_\alpha}^* \lambda^* - \boldsymbol{F}_{B_{t\alpha}}^{\text{hll}}}{B_x}, \quad (43)
$$

which may become ill defined as $B_x \to 0$. However this is not the case here, since what really matters in the induction and energy fluxes in Eq. (40) is $B_x \boldsymbol{v}_{t_\alpha}^*$ rather than the transverse velocity. Finally, the energy can be obtained from Eq. (39) by using either $\boldsymbol{v}^* \cdot \boldsymbol{B}^*$ or $\boldsymbol{m}^* \cdot \boldsymbol{B}^*/\rho^*$.

The Multi state HLL Solver: HLLD

The formulation adopted by Miyoshi and Kusano [35] solves the apparent incompatibilities introduced by the HLLC approach. The "HLLD solver" ("D" stands for "discontinuity"), introduced by [35], approximates the structure of the Riemann fan introducing five waves corresponding to two outermost fast shocks (λ_L and λ_R) and two rotational discontinuities (λ_L^* and λ_R^*) separated by a contact mode in the middle (λ_c^*). The resulting structure comprises four states, $\boldsymbol{U}_L^*$, $\boldsymbol{U}_L^{**}$, $\boldsymbol{U}_R^{**}$, and $\boldsymbol{U}_R^*$ as shown in Fig. 6. Across the rotational waves, density, total pressure, and the normal component of velocity remain

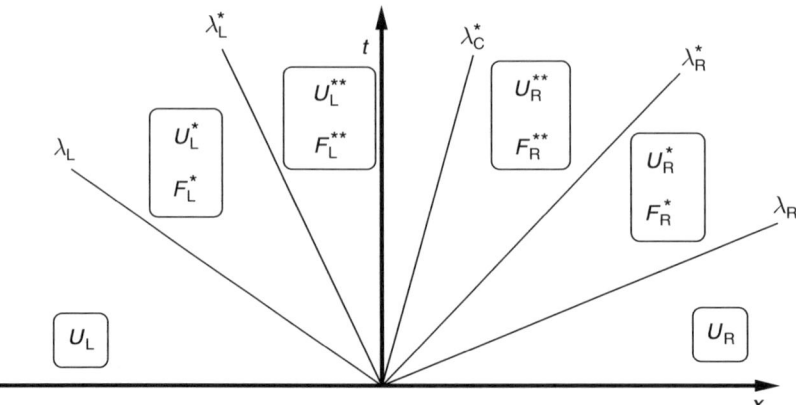

Fig. 6. Approximate structure to the Riemann fan used by the HLLD solver. Five waves separating four states are adopted. The *outermost waves* λ_L and λ_R correspond to fast shocks, while λ_L^* and λ_R^* identify rotational discontinuity. The *central wave* λ_c^* is the entropy mode

continuous. On the other hand, one must conserve $\boldsymbol{B}_t$, $\boldsymbol{v}$, and total pressure p through the contact mode. Since slow modes are not allowed inside the structure, p and u can be assumed constant throughout the Riemann fan and the normal velocity corresponds to the speed of the middle (entropy) wave. State and flux vectors retain the representation given by Eq. (36) in terms of eight unknown quantities in each state: $\rho^*, \lambda^*, \boldsymbol{v}_t^*, \boldsymbol{B}_t^*, E^*$, and p^*. As shown below, this formulation leads to a well-posed problem.

As a starting point, one can conveniently extend the derivation of the consistency conditions (34) and (35) to

$$\frac{(\lambda_R - \lambda_R^*)\,\boldsymbol{U}_R^* + (\lambda_R^* - \lambda_c^*)\,\boldsymbol{U}_R^{**} + (\lambda_c^* - \lambda_L^*)\,\boldsymbol{U}_L^{**} + (\lambda_L^* - \lambda_L)\,\boldsymbol{U}_L^*}{\lambda_R - \lambda_L} = \boldsymbol{U}^{\mathrm{hll}},$$
(44)

and, setting $\eta = 1/\lambda$,

$$\frac{(\eta_R - \eta_R^*)\,\boldsymbol{F}_R^* + (\eta_R^* - \eta_c^*)\,\boldsymbol{F}_R^{**} + (\eta_c^* - \eta_L^*)\,\boldsymbol{F}_L^{**} + (\eta_L^* - \eta_L)\,\boldsymbol{F}_L^*}{\eta_R - \eta_L} = \boldsymbol{F}^{\mathrm{hll}}.$$
(45)

Since $\rho_\alpha^* = \rho_\alpha^{**}$, the density and momentum components of (44) yield the obvious choice

$$\lambda_c^* \equiv u_L^* = u_L^{**} = u_R^{**} = u_R^* = \frac{m_x^{\mathrm{hll}}}{\rho^{\mathrm{hll}}}.$$
(46)

Likewise, using the same components in Eq. (44) one has

$$p^* \equiv p_L^* = p_L^{**} = p_R^{**} = p_R^* = F_{[m_x]}^{\mathrm{hll}} + B_x^2 - F_{[\rho]}^{\mathrm{hll}}\lambda_c^*.$$
(47)

The jump conditions across the outermost waves can now be used to determine the flow variables in the *star* regions as

$$\rho_\alpha^* = \rho_\alpha \frac{\lambda_\alpha - u_\alpha}{\lambda_\alpha - \lambda_c^*}, \tag{48}$$

$$\boldsymbol{B}_{t_\alpha}^* = \boldsymbol{B}_{t_\alpha} \frac{\rho_\alpha (\lambda_\alpha - u_\alpha)^2 - B_x^2}{\rho_\alpha (\lambda_\alpha - u_\alpha)(\lambda_\alpha - \lambda_c^*) - B_x^2}, \tag{49}$$

$$\boldsymbol{v}_{t_\alpha}^* = \boldsymbol{v}_{t_\alpha} - B_x \frac{\boldsymbol{B}_{t_\alpha}^* - \boldsymbol{B}_{t_\alpha}}{\rho_\alpha (\lambda_\alpha - u_\alpha)}, \tag{50}$$

$$E_\alpha^* = \frac{(\lambda_\alpha - u_\alpha) E_\alpha - p_\alpha u_\alpha + p^* \lambda_c^* + B_x (\boldsymbol{v}_\alpha \cdot \boldsymbol{B}_\alpha - \boldsymbol{v}_\alpha^* \cdot \boldsymbol{B}_\alpha^*)}{\lambda_\alpha - \lambda_c^*}, \tag{51}$$

where $\alpha = L$ or $\alpha = R$ for the left or right state, respectively.

The jump conditions across the rotational modes λ_L^* and λ_R^*,

$$\lambda_\alpha^* (\boldsymbol{U}_\alpha^{**} - \boldsymbol{U}_\alpha^*) = \boldsymbol{F}_\alpha^{**} - \boldsymbol{F}_\alpha, \tag{52}$$

cannot be solved independently to satisfy $\boldsymbol{B}_{tL}^{**} = \boldsymbol{B}_{tR}^{**}$ and $\boldsymbol{v}_{tL}^{**} = \boldsymbol{v}_{tR}^{**}$, unless the equations are linearly dependent. Indeed, only two out of the four relations for the transverse vector fields (for each wave) can be regarded independent, provided the speeds of the discontinuities are chosen to satisfy

$$\lambda_L^* = \lambda_c^* - \frac{|B_x|}{\sqrt{\rho_L^*}}, \quad \lambda_R^* = \lambda_c^* + \frac{|B_x|}{\sqrt{\rho_R^*}}. \tag{53}$$

The consistency condition (44) can now be used to find the state variables on either side of the entropy wave:

$$\boldsymbol{v}_{tL}^{**} = \boldsymbol{v}_{tR}^{**} = \frac{\sqrt{\rho_L^*}\boldsymbol{v}_{tL}^* + \sqrt{\rho_R^*}\boldsymbol{v}_{tR}^* + (\boldsymbol{B}_{tR} - \boldsymbol{B}_{tL})\sigma_x}{\sqrt{\rho_L^*} + \sqrt{\rho_R^*}}, \tag{54}$$

$$\boldsymbol{B}_{tL}^{**} = \boldsymbol{B}_{tR}^{**} = \frac{\sqrt{\rho_L^*}\boldsymbol{B}_{tR}^* + \sqrt{\rho_R^*}\boldsymbol{B}_{tL}^* + \sqrt{\rho_L^* \rho_R^*}(\boldsymbol{v}_{tR} - \boldsymbol{v}_{tL})\sigma_x}{\sqrt{\rho_R^*} + \sqrt{\rho_L^*}}, \tag{55}$$

with $\sigma_x = \text{sign}(B_x)$. For the energy one finds

$$E_L^{**} = E_L^* - \sqrt{\rho_L^*}(\boldsymbol{v}_L^* \cdot \boldsymbol{B}_L^* - \boldsymbol{v}_L^{**} \cdot \boldsymbol{B}_L^{**})\sigma_x, \tag{56}$$

$$E_R^{**} = E_R^* + \sqrt{\rho_R^*}(\boldsymbol{v}_R^* \cdot \boldsymbol{B}_R^* - \boldsymbol{v}_R^{**} \cdot \boldsymbol{B}_R^{**})\sigma_x. \tag{57}$$

As one can see, the problem is defined in terms of 32 unknowns (eight per state) complemented by 32 independent non trivial equations: 14 across the outermost waves (seven for $\alpha = L$ and seven for $\alpha = R$), six continuity conditions across the contact mode:

$$[p] = [u] = 0, \quad [\boldsymbol{B}_t] = [\boldsymbol{v}_t] = \boldsymbol{0}, \tag{58}$$

and 12 conditions across the rotational waves, six of which are given by

$$[\rho] = [u] = [p] = 0 , \qquad (59)$$

(across λ_L^* and λ_R^*), plus six *independent* jump conditions for B_t, v_t, and E. It is now straightforward to compute the inter cell numerical flux:

$$
\hat{f} = \begin{cases}
F_L & \text{if} \quad \lambda_L \geq 0, \\
F_L^* & \text{if} \quad \lambda_L \leq 0 \leq \lambda_L^* , \\
F_L^{**} & \text{if} \quad \lambda_L^* \leq 0 \leq \lambda_c^* , \\
F_R^{**} & \text{if} \quad \lambda_c^* \leq 0 \leq \lambda_R^* , \\
F_R^* & \text{if} \quad \lambda_R^* \leq 0 \leq \lambda_R , \\
F_R & \text{if} \quad \lambda_R \leq 0 .
\end{cases} \qquad (60)
$$

The HLLD flux is found to be robust with an accuracy comparable to that of the Roe scheme. Recently, [33], this methodology has been extended to the isothermal MHD equations.

The Scheme of Roe

Roe's scheme [39] tries to resolve the initial jump by replacing the original conservation law with a linearized system of constant coefficients. Thus one proceeds by seeking the exact solution to the following linearized Riemann problem:

$$
\begin{cases}
\dfrac{\partial U}{\partial t} + \overline{A}_{LR} \cdot \dfrac{\partial U}{\partial x} = 0, \\
U(x,0) = \begin{cases} U_L & \text{for} \quad x < 0, \\ U_R & \text{for} \quad x > 0. \end{cases}
\end{cases} \qquad (61)
$$

where $\overline{A}_{RL} \equiv \overline{A}(U_L, U_R)$ is a constant matrix. The matrix $\overline{A}_{LR}$ is called a Roe matrix if the following requirements are met [46]:

- Consistency with the original conservation law: $\overline{A}(U, U) = A(U)$, where A is simply the jacobian of the flux, i.e., $A(U) = \partial F(U)/\partial U$
- Conservation across discontinuities:

$$F(U_R) - F(U_L) = \overline{A}_{LR} \cdot (U_R - U_L) . \qquad (62)$$

- Hyperbolicity: $\overline{A}$ must have a complete set of real eigenvalues $\lambda_k(U_L, U_R)$ and associated left and right eigenvectors L_k and R_k such that

$$A \cdot R_k = \lambda_k R_k, \quad L_k \cdot A = \lambda_k L_k. \qquad (63)$$

Roe's scheme is thus equivalent to the solution of the Riemann problem for a system of advection equation with linear constant coefficients, for which the flux function takes the form:

$$\hat{f} = \frac{F_L + F_R}{2} - \frac{1}{2} \sum_k (L_k \cdot \Delta U) |\lambda_k| R_k, \qquad (64)$$

and $\Delta U = U_R - U_L$. Despite the original method of Roe, initially developed for the Euler equation of gasdynamics, has been extended to MHD using simpler averages (e.g., the arithmetic average, see for example [5, 37, 42, 44], Cargo and Gallice [6] showed how the correct Roe matrix for MHD can be constructed for any value of the specific heat ratio Γ. For the sake of completeness, we report hereafter the expressions. To this purpose, we introduce two different Roe averages for any flow quantity Q by defining

$$\overline{Q} = \frac{\sqrt{\rho_L}Q_L + \sqrt{\rho_R}Q_R}{\sqrt{\rho_L} + \sqrt{\rho_R}} , \quad \underline{Q} = \frac{\sqrt{\rho_R}Q_L + \sqrt{\rho_L}Q_R}{\sqrt{\rho_L} + \sqrt{\rho_R}} . \qquad (65)$$

Velocity and specific enthalpy $H = (E + p)/\rho$ are computed using the first definition, whereas density and transverse components of the field follow the second one. In other words, $v, H \in \overline{Q}$, while $\rho, B_t \in \underline{Q}$. Next we compute the jumps in conservative variables, that is, $\Delta U = U_R - U_L$. The pressure jump is derived accordingly as

$$\Delta p = (\Gamma - 1) \left[\left(\frac{v^2}{2} - X \right) \Delta \rho - v \cdot \Delta(\rho v) + \Delta E - B \cdot \Delta B \right], \qquad (66)$$

with

$$X = \frac{\Delta B \cdot \Delta B}{2 \left(\sqrt{\rho_L} + \sqrt{\rho_R} \right)^2} . \qquad (67)$$

Now, according to the chosen set of variables, we compute the sound speed a and Alfvén velocity c_A as

$$a^2 = (2 - \Gamma)X + (\Gamma - 1) \left(H - \frac{v^2}{2} - \frac{B^2}{\rho} \right), \quad c_A^2 = \frac{B_x^2}{\rho} , \qquad (68)$$

while fast and slow velocities c_f and c_s are given by

$$c_f^2 = \frac{1}{2} \left(a^2 + \frac{B^2}{\rho} + \sqrt{(a^2 - b^2)^2 + 4b_t^2 a^2} \right), \quad c_s^2 = \frac{a^2 c_A^2}{c_f^2} . \qquad (69)$$

Having defined the Roe average, we now give the expressions for the right eigenvectors R_k and wave strengths $L_k \cdot \Delta U$ required in Eq. (64). For convenience, we define [6]

$$\alpha_f = \frac{a^2 - c_s^2}{c_f^2 - c_s^2} , \quad \alpha_s = \frac{c_f^2 - a^2}{c_f^2 - c_s^2} , \quad S_s = \alpha_s c_s \sigma_x , \quad S_f = \alpha_f c_f \sigma_x , \qquad (70)$$

where $\sigma_x = \text{sign}(B_x)$. The right eigenvectors for the fast modes $f\pm$ with eigenvalues $\lambda_{f\pm} = u \pm c_f$ are

$$\boldsymbol{R}_{f\pm} = \begin{pmatrix} \alpha_f \\ \alpha_f \lambda_{f\pm} \\ (\alpha_f \boldsymbol{v}_t \mp S_s \boldsymbol{\beta}_t) \\ \dfrac{\alpha_s}{\sqrt{\rho}} a \boldsymbol{\beta}_t \\ \alpha_f \left(H^* - \dfrac{\boldsymbol{B}^2}{\rho} \pm v_x c_f \right) \mp S_s \left(\boldsymbol{v}_t \cdot \boldsymbol{\beta}_t \right) + \dfrac{\alpha_s}{\sqrt{\rho}} a \left| \boldsymbol{B}_t \right| \end{pmatrix} . \tag{71}$$

Similarly, the expressions for the slow modes $s\pm$ associated with $\lambda_{s\pm} = u \pm c_s$ follow:

$$\boldsymbol{R}_{s\pm} = \begin{pmatrix} \alpha_s \\ \alpha_s \lambda_{s\pm} \\ (\alpha_s \boldsymbol{v}_t \pm S_f \boldsymbol{\beta}_t) \\ -\dfrac{\alpha_f}{\sqrt{\rho}} a \boldsymbol{\beta}_t \\ \alpha_s \left(H^* - \dfrac{\boldsymbol{B}^2}{\rho} \pm v_x c_s \right) \pm S_f \left(\boldsymbol{v}_t \cdot \boldsymbol{\beta}_t \right) - \dfrac{\alpha_f}{\sqrt{\rho}} a \left| \boldsymbol{B}_t \right| \end{pmatrix} . \tag{72}$$

The linear modes are given by the rotational Alfvén waves propagating at $\lambda_{A\pm} = u \pm c_A$ and the entropy contact mode moving at the flow velocity $\lambda_u = u$. The associated expressions for the corresponding right eigenvectors are found to be

$$\boldsymbol{R}_u = \begin{pmatrix} 1 \\ u \\ \boldsymbol{v}_t \\ \boldsymbol{0} \\ \dfrac{\boldsymbol{v}^2}{2} + \dfrac{\Gamma - 2}{\Gamma - 1} X \end{pmatrix} , \quad \boldsymbol{R}_{A\pm} = \begin{pmatrix} 0 \\ 0 \\ \pm \rho \boldsymbol{\beta}_t \times \hat{\boldsymbol{n}} \\ -S\sqrt{\rho} \boldsymbol{\beta}_t \times \hat{\boldsymbol{n}} \\ \pm \rho \left(\boldsymbol{v}_t \times \boldsymbol{\beta}_t \cdot \hat{\boldsymbol{n}} \right) \end{pmatrix} . \tag{73}$$

We conclude the presentation by providing the expressions for the wave strengths:

$$\boldsymbol{L}_{f\pm} \cdot \Delta \boldsymbol{U} = \frac{\alpha_f Y \mp \rho S_s \boldsymbol{\beta}_t \cdot \Delta \boldsymbol{v}_t \pm \rho \alpha_f c_f \Delta u + \sqrt{\rho} \alpha_s a \boldsymbol{\beta}_t \cdot \Delta \boldsymbol{B}_t}{2a^2} , \tag{74}$$

$$\boldsymbol{L}_{s\pm} \cdot \Delta \boldsymbol{U} = \frac{\alpha_s Y \pm \rho S_f \boldsymbol{\beta}_t \cdot \Delta \boldsymbol{v}_t \pm \rho \alpha_s c_s \Delta u - \sqrt{\rho} \alpha_f a \boldsymbol{\beta}_t \cdot \Delta \boldsymbol{B}_t}{2a^2} , \tag{75}$$

and

$$\boldsymbol{L}_{A\pm} \cdot \Delta \boldsymbol{U} = \frac{\boldsymbol{\beta}_t}{2} \times \left[\mp \Delta \boldsymbol{v} + \frac{S}{\sqrt{\rho}} \Delta \boldsymbol{B}_t \right] \cdot \hat{\boldsymbol{n}} , \quad \boldsymbol{L}_u \cdot \Delta \boldsymbol{U} = 1 - \frac{Y}{a^2} , \tag{76}$$

where $Y = (X \Delta \rho + \Delta p)$, and $\hat{\boldsymbol{n}}$ is the unit vector normal to the discontinuity front.

The Riemann solver of Roe is able to correctly capture any isolated regular discontinuity, including fast, slow, Alfvén, and contact fronts. In some circumstances, however, the Roe solver can fail giving rise to unphysical (negative) pressures or trigger spurious numerical instabilities, such as the *carbuncle* phenomenon, [38]. Also, the assumption of strict hyperbolicity ceases when some of the eigenvectors are not linearly independent anymore, [40]. For this reason, regions of the flow that might potentially cause problems are treated with an hybrid Riemann solver that selectively switch from Roe's scheme to a more robust (albeit less accurate) scheme such as HLL.

4 The $\nabla \cdot \boldsymbol{B} = 0$ Condition

The absence of magnetic monopoles is mathematically expressed by Gauss' law for magnetism, i.e.,

$$\nabla \cdot \boldsymbol{B} = 0 \,, \tag{77}$$

which simply states that, for any control volume, the net magnetic flux across the boundary is identically zero. In other words, magnetic fields do not have sink or sources. The solenoidal condition is not an evolutionary equation but, rather, a constraint to be fulfilled at all times. Indeed, if $\nabla \cdot \boldsymbol{B} = 0$ at some initial time, then by taking the divergence of Faraday's law of electromagnetism

$$\frac{\partial \boldsymbol{B}}{\partial t} + \nabla \times \boldsymbol{\Omega} = 0 \,, \tag{78}$$

one has $\partial \nabla \cdot \boldsymbol{B}/\partial t = 0$, i.e., the field is divergence-free at all times.

From a numerical point of view, however, this condition is fulfilled only at the truncation level and non solenoidal components may be generated during the evolution. This causes unphysical accelerations of the plasma in the direction parallel to the field lines, as outlined by [4]. Generally speaking, it is not possible for a numerical scheme to satisfy $\nabla \cdot \boldsymbol{B} = 0$ for any type of discretization. Thus different method of solutions can be sought and the robustness of one strategy over another can be established on a practical base by extensive numerical testing. In what follows, we give a rather concise description of the different approaches embraced up to now in Godunov type schemes, and follow a two-category classification.

In the first one, a cell-centered representation of the magnetic field is used. This naturally extends the formalism developed for the Euler equation of gasdynamics to MHD and offers the advantage of being conceptually simple and easy to implement in existing hydrocodes. Moreover, exploiting a cell-centered representation of all conserved quantities makes the extension to adaptive and unstructured grids straightforward.

In the second class, the magnetic field has a staggered representation whereby field components live on the face they are normal to. Hydrodynamics variables (density, velocity, and pressure) retains their usual collocation at the cell center. This provides a framework by which the induction equation (78) is more naturally updated using Stoke's theorem and the divergence-free condition is fulfilled to machine accuracy, if $\nabla \cdot \boldsymbol{B} = 0$ initially.

4.1 Cell-Centered Methods

Powell's 8-Wave Formulation

In the 8-wave formulation [36], Gauss' law for magnetism is discarded in the physical derivation of the MHD equations [37]. From the vector identity

$$\nabla \times (\boldsymbol{v} \times \boldsymbol{B}) = \boldsymbol{v} (\nabla \cdot \boldsymbol{B}) - \boldsymbol{B} (\nabla \cdot \boldsymbol{v}) + (\boldsymbol{v} \cdot \nabla) \boldsymbol{B} - (\boldsymbol{v} \cdot \nabla) \boldsymbol{B} \,, \tag{79}$$

and the fact that $\nabla \cdot (\boldsymbol{v}\boldsymbol{B}) = (\nabla \cdot \boldsymbol{v})\boldsymbol{B} + (\boldsymbol{v} \cdot \nabla)\boldsymbol{B}$, one can re write Fraday's law as

$$\frac{\partial \boldsymbol{B}}{\partial t} + \nabla \cdot (\boldsymbol{v}\boldsymbol{B} - \boldsymbol{B}\boldsymbol{v}) = -\boldsymbol{v} (\nabla \cdot \boldsymbol{B}) \,. \tag{80}$$

Likewise, applying the same arguments to the momentum equation

$$\frac{\partial (\rho \boldsymbol{v})}{\partial t} + \nabla \cdot [\rho \boldsymbol{v}\boldsymbol{v} + p_g \boldsymbol{I}] = \boldsymbol{j} \times \boldsymbol{B} \quad \text{with} \quad \boldsymbol{j} = \nabla \times \boldsymbol{B} \,, \tag{81}$$

and to the energy equation

$$\frac{\partial}{\partial t} \left(\rho \epsilon + \rho \frac{v^2}{2} \right) + \nabla \cdot \left[\left(\rho \epsilon + p_g + \rho \frac{v^2}{2} \right) \boldsymbol{v} \right] = \boldsymbol{j} \cdot \boldsymbol{B} \,, \tag{82}$$

the divergence form of the equation takes the form [36, 37]

$$\frac{\partial}{\partial t} \begin{pmatrix} \rho \\ \rho \boldsymbol{v} \\ \boldsymbol{B} \\ E \end{pmatrix} + \nabla \cdot \begin{bmatrix} \rho \boldsymbol{v} \\ \rho \boldsymbol{v}\boldsymbol{v} - \boldsymbol{B}\boldsymbol{B} + p\boldsymbol{I} \\ \boldsymbol{v}\boldsymbol{B} - \boldsymbol{B}\boldsymbol{v} \\ (E + p) \boldsymbol{v} - (\boldsymbol{v} \cdot \boldsymbol{B}) \boldsymbol{B} \end{bmatrix} = -\nabla \cdot \boldsymbol{B} \begin{pmatrix} 0 \\ \boldsymbol{B} \\ \boldsymbol{v} \\ \boldsymbol{v} \cdot \boldsymbol{B} \end{pmatrix} \,, \tag{83}$$

where $p = p_g + \boldsymbol{B}^2/2$ is the total (thermal + magnetic) pressure and $E = \rho \epsilon + \rho v^2/2 + \boldsymbol{B}^2/2$ is the total energy density.

Although the source term should be physically zero, Powell showed that its inclusion changes the character of the equations by introducing an additional eighth wave corresponding to the propagation of jumps in the component of magnetic field normal (B_x) to a given interface. The other seven waves are left unchanged from the traditional formulation, since it can be shown that none of them carries a jump in B_x.

The 8-wave formulation leads to a symmetrizable form of conservation laws which, among other properties [20, 23], makes the system Galilean invariant. This property does not hold anymore if the source term is dropped. Powell also showed that the 8-wave formulation leads to the passive advection of $(\nabla \cdot \boldsymbol{B})/\rho$. The latter property states that magnetic monopoles are advected by the flow as they are created. However, $\nabla \cdot \boldsymbol{B} = 0$ is not satisfied in any particular discretization, but only to truncation level.

Powell's method has the advantage of being computationally inexpensive and easy to implement without introducing significative complexities. Numerical tests show that this modification, in any existing shock-capturing MHD code, results in stable and robust schemes. On the other hand, the 8-wave formulation leads to a non conservative form of the equations. Although Powell claimed that deviations from the conservation should be very small, in [48] Tóth proved that near discontinuities (shock waves) large conservation errors may be produced.

Divergence Cleaning

In [12], the divergence free constraint is enforced by solving a modified system of conservation laws, where the induction equation is coupled to a generalized Lagrange multiplier. According to this modification, the induction equation is replaced with

$$\frac{\partial \boldsymbol{B}}{\partial t} + \nabla \cdot (\boldsymbol{v}\boldsymbol{B} - \boldsymbol{B}\boldsymbol{v}) + \nabla \psi = 0 \,, \tag{84}$$

and the solenoidal condition is expressed through

$$\mathcal{D}(\psi) + \nabla \cdot \boldsymbol{B} = 0 \,, \tag{85}$$

where $\mathcal{D}$ is a differential operator. For any choice of $\mathcal{D}$, one can show that the divergence of $\boldsymbol{B}$ and the scalar function ψ satisfy the same equation, namely

$$\frac{\partial}{\partial t}\mathcal{D}\left(\nabla \cdot \boldsymbol{B}\right) - \triangle\left(\nabla \cdot \boldsymbol{B}\right) = 0 \quad \Longleftrightarrow \quad \frac{\partial \mathcal{D}(\psi)}{\partial t} - \triangle\psi = 0 \,. \tag{86}$$

An *elliptic* correction results from taking $\mathcal{D} = 0$. This is equivalent to the projection method introduced by [4] and explained in the next section.

By taking $\mathcal{D}(\psi) = \psi/c_p^2$ with $c_p > 0$, one ends up with a *parabolic* correction. Simple manipulations show that, in this case, ψ can be trivially eliminated from the equations and the induction equation becomes

$$\frac{\partial \boldsymbol{B}}{\partial t} + \nabla \cdot (\boldsymbol{v}\boldsymbol{B} - \boldsymbol{B}\boldsymbol{v}) = c_p^2 \nabla\left(\nabla \cdot \boldsymbol{B}\right) \,, \tag{87}$$

which states that local divergence errors are damped by an additional dissipation mechanism, provided compatible boundary conditions are used.

A third, *hyperbolic* correction follows if $\mathcal{D}(\psi) = c_h^{-2}\partial\psi/\partial t$ is chosen, with $c_h > 0$. The hyperbolic correction propagates local divergence errors to the boundary with the finite speed c_h.

Finally, a mixed hyperbolic/parabolic correction can be prescribed by summing up the respective differential operators. The resulting divergence constrain becomes

$$\frac{\partial \psi}{\partial t} + c_h^2 \nabla \cdot \boldsymbol{B} = -\frac{c_h^2}{c_p^2} \psi \,, \tag{88}$$

offering both dissipation and propagation of divergence errors. In the mixed formulation, divergence errors are transported to the domain boundaries with the maximal admissible speed and are damped at the same time.

The main advantage of the approaches outlined above and, in particular of the mixed formulation, is to preserve the full conservation form of the original MHD system. Indeed, only the equation for the unphysical scalar function ψ contains a source term. In addition, divergence errors are transported by two waves with speeds independent of the fluid velocity. In this respect, this procedure may be viewed as an extension of Powell's divergence wave. Finally, by taking advantage of operator splitting techniques, the equation for the normal component of the magnetic field and ψ are decoupled from the other equations. This allows for the solution of a 2×2 linear system of hyperbolic equations, thus considerably reducing the computational effort.

Projection Scheme

The projection scheme, originally proposed by Brackbill and Barnes [4], consists in applying a correction step to the magnetic field $\boldsymbol{B}^*$ obtained after the base scheme evolution (i.e., without any correction). In general, since the base scheme will not preserve the divergence free condition, one can use Helmholtz decomposition to resolve $\boldsymbol{B}^*$ as the sum of an irrotational and a solenoidal vector field, associated with scalar and vector potentials φ and $\boldsymbol{A}$, respectively:

$$\boldsymbol{B}^* = \nabla \varphi + \nabla \times \boldsymbol{A} \,. \tag{89}$$

The physically relevant part of the field is the one associated with the vector potential, i.e., $\nabla \times \boldsymbol{A}$. The correct divergence-free magnetic field is recovered by subtracting the unphysical contribution coming from the irrotational component, i.e.,

$$\boldsymbol{B}^{\mathrm{new}} = \boldsymbol{B}^* - \nabla \varphi \,, \tag{90}$$

after the Poisson equation

$$\nabla^2 \varphi = \nabla \cdot \boldsymbol{B}^* \,, \tag{91}$$

obtained by taking the divergence of Eq. (89), has been solved. As noticed by Tóth [48], the difference operators approximating the divergence and gradient in Eqs. (90) and (91) must be consistently used to compute the Laplacian operator in the Poisson equation. He also proved that the correction given by Eq. (91) changes the solution from the base scheme $\boldsymbol{B}^*$ to the closest divergence-free discrete representation of the field, by introducing the smallest possible correction. Furthermore, the method provides a consistent discretization without reducing the order of accuracy of the base scheme, even in presence of discontinuities.

Of course, the price one has to pay is the solution of a Poisson equation which can be rather expensive, especially on parallel computers or on adaptive grids. Under many circumstances, however, it may not be necessary to solve Eq. (91) to machine accuracy, and divergence errors may as well be kept below some predefined tolerance, typically a small fraction of the $|\nabla \cdot \boldsymbol{B}^*|$ generated in a single time step from the base scheme. This makes iterative solvers, like the conjugate gradient type linear solvers [24, 49], particularly efficient and flexible candidates to solve Eq. (91).

4.2 Constrained Transport

In the constrained transport method (CT henceforth), originally devised by [17], the induction equation is discretized by adopting a staggered representation of magnetic and electric vector fields. This formulation is better understood by integrating Faraday's law on a given surface S bounded by the curve ∂S and using Stoke's theorem to obtain a surface integration:

$$\frac{d}{dt} \int_S \boldsymbol{B} \cdot d\boldsymbol{S} = - \int_{\partial S} \boldsymbol{\Omega} \cdot d\boldsymbol{l} \,, \tag{92}$$

where $\boldsymbol{\Omega}$ is the electromotive force. In this form, the three components of magnetic field are evolved on the zone face to which they are orthogonal and are treated as surface averages, see Fig. 7. Each component has, therefore, a different spatial collocation in the control volume. For a Cartesian cell with lower and upper coordinate limits given by (x_-, y_-, z_-) and (x_+, y_+, z_+), the surface averaged magnetic fields are

$$\bar{B}_x^{\pm} = \frac{1}{\Delta y \Delta z} \int \int B_x(x_{\pm}, y, z) \, dy dz \,, \tag{93}$$

$$\bar{B}_y^{\pm} = \frac{1}{\Delta x \Delta z} \int \int B_y(x, y_{\pm}, z) \, dx dz \,, \tag{94}$$

$$\bar{B}_z^{\pm} = \frac{1}{\Delta x \Delta y} \int \int B_z(x, y, z_{\pm}) \, dx dy \,, \tag{95}$$

where the integrals extend from the lower to the upper bounds of the cell and $\Delta x = x_+ - x_-$, $\Delta y = y_+ - y_-$, $\Delta z = z_+ - z_-$ are the zone widths. Note that the staggered collocation is perfectly consistent with the traditional 7-wave approach where the normal component of the field B_x is not allowed to have a jump, while the tangential components certainly can. From this perspective, B_x is regarded as a parameter when solving Riemann problems at zone interfaces and does not need to be reconstructed from the cell center, being already defined in the correct position. The discrete version of Eq. (92) reduces to

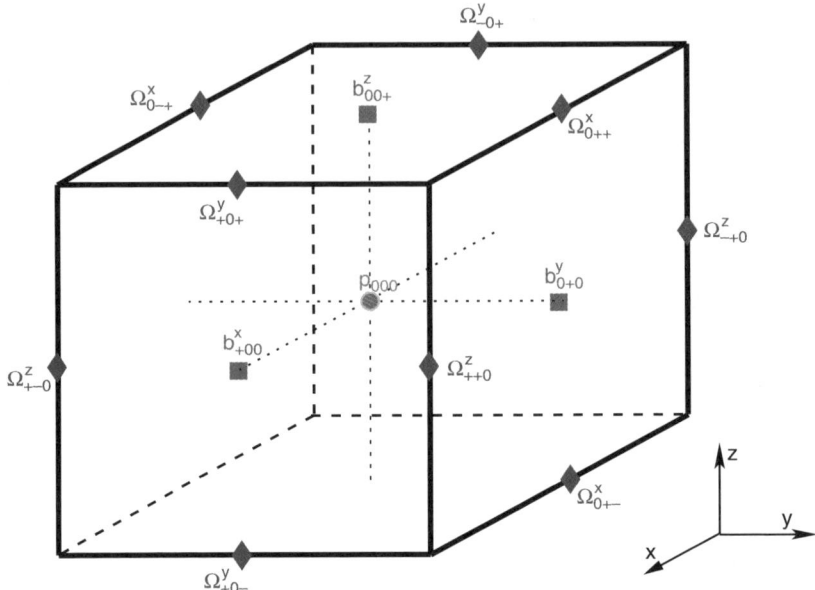

Fig. 7. Spatial collocation of flow variables in a 3D Cartesian cell. Cell-centered quantities (*circle*) include density, pressure, and velocity. Magnetic fields (*squares*) are face-centered, while electric fields (*diamonds*) are edge-centered

$$\frac{d\bar{B}_n^{\pm}}{dt} = -\sum_{m,l} \epsilon_{nml} \frac{\Delta_m \tilde{\Omega}_l^{\pm}}{\Delta h_m}, \tag{96}$$

where $\{n, m, l\} = x, y, z$ ($\Delta h_n = \Delta x, \Delta y, \Delta z$) and $\Delta_m \tilde{\Omega}_l$ is the difference between the line-averaged electric field

$$\tilde{\Omega}_l = \frac{1}{\Delta h_l} \int \Omega_l dh_l, \tag{97}$$

evaluated at opposite edges in the l-direction. In this formalism, one can verify that the divergence-free condition in its integral form,

$$\sum_n \frac{\bar{B}_n^+ - \bar{B}_n^-}{\Delta h_n} = 0, \tag{98}$$

is preserved to machine accuracy if the initial field has zero divergence in this discretization.

Note that the electric fields, Eq. (97), are evaluated as line integrals along the cell edges. This issue has been coped with in a number of different ways, starting with the earlier work of [3, 10, 11, 43], who incorporated the CT discretization in Godunov-type numerical schemes. Typical upwind schemes, in fact, achieve second-order accuracy by interpolating cell-centered values to the face midpoint and then solving a Riemann problem between the resulting left and right states. Electric fields are thus available either at the cell center by properly averaging velocity and magnetic vector fields [10, 11] or at the face center using upwinded fluxes [3] from the induction equation. Therefore, some kind of spatial interpolation is required to produce the edge-centered electromotive force. In the original approach of [3], for instance, a simple arithmetic averaging between the upwind fluxes coming the four faces adjacent to a cell edge is adopted, see Fig. 8. In 2-D, for simplicity, the edge-centered electric field $\Omega \equiv \Omega_z$ is produced as:

$$\tilde{\Omega}_{i+\frac{1}{2},j+\frac{1}{2}} = \frac{\hat{\Omega}_{i+\frac{1}{2},j} + \hat{\Omega}_{i+\frac{1}{2},j+1} + \hat{\Omega}_{i,j+\frac{1}{2}} + \hat{\Omega}_{i+1,j+\frac{1}{2}}}{4}. \tag{99}$$

where the $\hat{\Omega}$'s follow the solution of Riemann problems at the corresponding interfaces. Despite its simplicity and effectiveness, the proposed average does

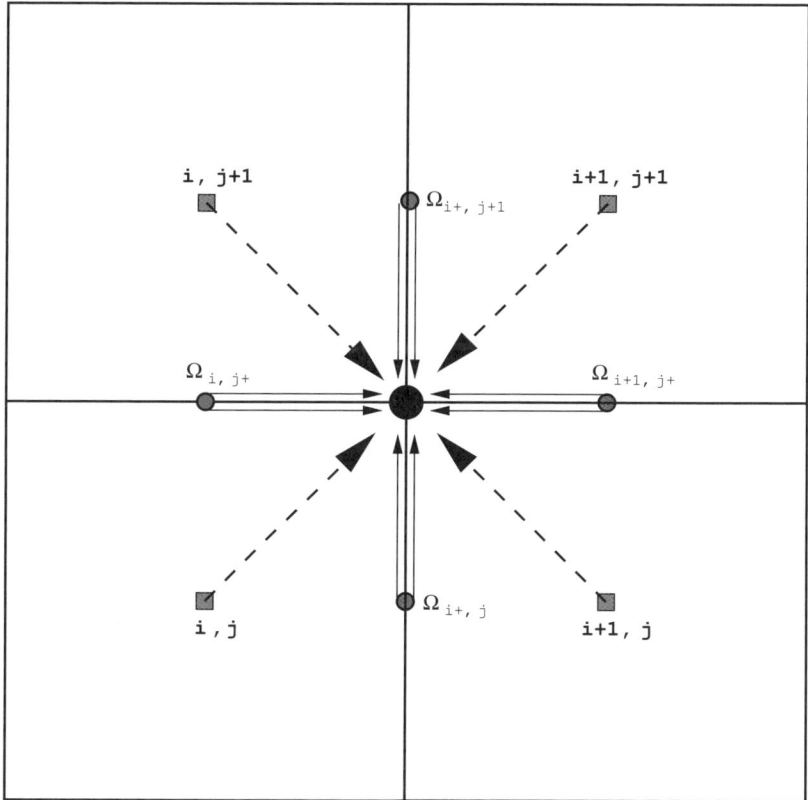

Fig. 8. Computation of the edge-centered electromotive force in 2D. The interpolation process can proceed by either averaging the vector fields available at the center (*squares* and *dashed arrows*) or the upwind fluxes available at zone interfaces (*circles* and *solid arrows*). In the picture $i+ = i + \frac{1}{2}$, $j+ = j + \frac{1}{2}$

not reduce to the equivalent solver for plane-parallel grid aligned flow. This failure can be traced back to the lack of a directional bias in the averaging formula [19]. This issue has been addressed by a number of investigators, at the same time offering alternative strategies to overcome this inconsistency.

In [29] and, more recently [30], this problem has been faced by noticing that the flux components appearing in the induction equation are defined in terms of point values at the intersections of cell faces, where different characteristic wave fans may overlap. An appropriate average of these flux components should follow a proper upwind selection rule, since a same flux component at the same collocation point results to have two independent representations in terms of characteristic wave fans. The resulting four-states combination cannot be reduced simply to an interpolation or averaging form based on the four cell-centered values of the arguments. Instead, specializing to the $\Omega \equiv \Omega_z$ flux, a single-valued numerical flux comes out by averaging over the two overlapping x and y Riemann wave fans at the $(x_{i+\frac{1}{2}}, y_{j+\frac{1}{2}})$ edge. This entails to a four-state flux function preserving the continuity and upwind properties along each direction:

$$\tilde{\Omega} = \langle \Omega \rangle - \phi_y + \phi_x \,, \tag{100}$$

where the first term $\langle \Omega \rangle$ expresses the smooth contribution, whereas ϕ_x and ϕ_y are the dissipative terms. For the HLL central-upwind scheme, for instance, smooth and dissipative terms take the form

$$\langle \Omega \rangle = \frac{\alpha_x^+ \alpha_y^+ \Omega^{LL} + \alpha_x^+ \alpha_y^- \Omega^{LR} + \alpha_x^- \alpha_y^+ \Omega^{RL} + \alpha_x^- \alpha_y^- \Omega^{RR}}{(\alpha_x^+ + \alpha_x^-)(\alpha_y^+ + \alpha_y^-)} \,, \tag{101}$$

and

$$\phi_x = \frac{\alpha_x^+ \alpha_x^-}{\alpha_x^+ + \alpha_x^-} \left(B_{y,i+1,j+\frac{1}{2}}^R - B_{y,i,j+\frac{1}{2}}^L \right) \,, \tag{102}$$

$$\phi_y = \frac{\alpha_y^+ \alpha_y^-}{\alpha_y^+ + \alpha_y^-} \left(B_{x,i+\frac{1}{2},j+1}^R - B_{x,i+\frac{1}{2},j}^L \right) \,, \tag{103}$$

where left (L) and right (R) superscripts give the corresponding interpolated point values with respect to the edge $(i + \frac{1}{2}, j + \frac{1}{2})$, see Fig. 9. In Eqs. (101), (102) and (103), $\alpha_x^\pm$ and $\alpha_y^\pm$ determine the opening of the x and y Riemann fans, in terms of estimates to the maximum $(+)$ and minimum $(-)$ characteristic velocities, respectively. Adopting the notations introduced in Eq. (29), one may take, for example

$$\alpha_x^\pm = \max \left(\pm \lambda_{S,i+\frac{1}{2},j}^x, \pm \lambda_{S,i+\frac{1}{2},j+1}^x, 0 \right) \tag{104}$$

where $S = R$ $(S = L)$ for α_x^+ (α_x^-). Similar considerations hold for $\alpha_y^\pm$. The constructed numerical flux is consistent with the fact that each component of the induction equation, for a given velocity field and expressed in terms of the vector potential $\mathbf{A}$, has the form of a scalar Hamilton–Jacobi equation. This

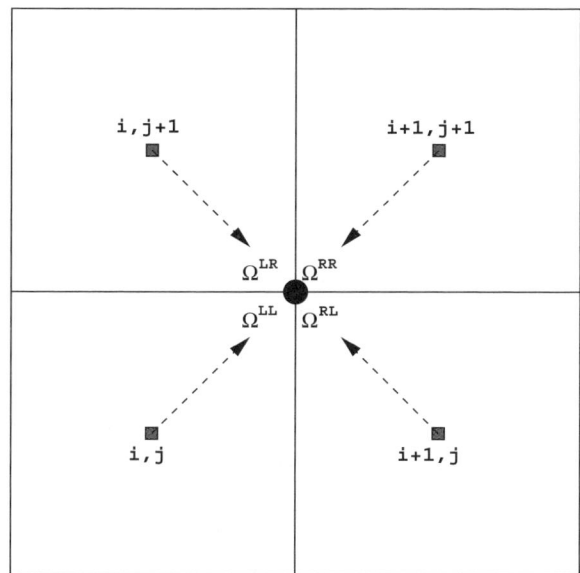

Fig. 9. Computation of the edge-centered electromotive force using the Upwind Constrained Transport (UCT) method of [30]. The four electric fields $\Omega^{LL}, \Omega^{LR}, \Omega^{RL}$, and Ω^{RR} are reconstructed from the cell centers using two-dimensional interpolation. The edge-centered electromotive force comes by averaging the two overlapping Riemann fan at $i + \frac{1}{2}, j + \frac{1}{2}$

formalism has been successfully extended to special and general relativistic flows in [13, 14].

A different, somewhat more empirical strategy is proposed by [19], where the authors construct a more elaborate spatial integration procedure for the electric field. The reconstruction rest upon an upwind selection rule depending on the sign of the contact mode, in a way specifically designed to properly control the amount of numerical dissipation.

In [51], the induction equation is modified by introducing an additional term containing a mixed (space–time mixed) second derivative of the electric field. The sign of this extra term is chosen to control the anti dissipative effect arising from the equivalent modified equation.

References

1. Balsara, D.S., ApJS, **116**, 133 (1998)
2. Balsara, D.S., & Spicer, D.S., JCP, **149**, 270 (1999)
3. Batten, P., Clarke, N., Lambert, C., & Causon, D.M., SIAM JSCom, **18**, 1553 (1997)
4. BrackBill, J.U., & Barnes, D.C., JCP, **35**, 426 (1980)
5. Brio, M., & Wu, C.C., JCP, **75**, 400 (1988)

6. Cargo, P., & Gallice, G., JCP, **136**, 446 (1997)
7. Dai, W., & Woodward, P.R., JCP, **111**, 354 (1994)
8. Dai, W., & Woodward, P.R., JCP, **115**, 485 (1994)
9. Dai, W., & Woodward, P.R., SIAM JSCom, **18**, 957 (1997)
10. Dai, W., & Woodward, P.R., ApJ, **494**, 317 (1998)
11. Dai W., & Woodward, P.R., JCP, **142**, 331 (1998)
12. Davis, S.F., SIAM JSSCom, **9**, 445 (1988)
13. Dedner, A., Kemm, F., Kröner, D., Munz, C.-D., Schnitzer, T., & Wesenberg M., JCP, **175**, 645 (2002)
14. Del Zanna, L., Bucciantini, N., & Londrillo, P., A & A, **400**, 397 (2003)
15. Del Zanna, L., Zanotti, O., Bucciantini, N., & Londrillo, P., A & A, **473**, 11 (2007)
16. Einfeldt, B., SIAM JNM, **25**, 294 (1988)
17. Evans, C. R., & Hawley J.F., JCP, **332**, 659 (1988)
18. Falle, S.A.E.G., Komissarov, S.S., & Joarder, P., MNRAS, **297**, 265 (1998)
19. Gardiner, T.A., & Stone, J.M., JCP, **205**, 509 (2005)
20. Godunov, S.K., DoKAN, SSSR, **139**, 521 (1972)
21. Gurski, K.F., SIAM JSCom, **25**, 2165 (2004)
22. Harten A., JCP, **49**, 357 (1983)
23. Harten, A., Lax, P.D., & van Leer, B., SIAMR, **25**, 35 (1983)
24. Hestenes, M.R., & Stiefel, E., JRNBS, **49**, 409 (1954)
25. Jeffrey A., & Taniuti T., Non-linear wave propagation. Academic Press, New York, (1964)
26. Kennel, C.F., Blandford, R.D., & Coppi, P., JPlPh, **42**, 299 (1989)
27. Lax, P.D., CPAM, **VII**, 159 (1954)
28. Lee, D., An Unsplit Staggered Mesh Scheme For Multidimensional Magneto-hydrodynamics: A Staggered Dissipation-control Differencing Algorithm Ph.D. Thesis, University of Maryland (College Park, Md.) (2006)
29. Li, S., JCP, **203**, 344 (2005)
30. Londrillo, P., & Del Zanna, L., ApJ, **530**, 508 (2000)
31. Londrillo, P., & Del Zanna, L., JCP, **195**, 17 (2004)
32. Mignone, A., & Bodo, G., MNRAS, **364**, 126 (2005)
33. Mignone, A., & Bodo, G., MNRAS, **368**, 1040 (2006)
34. Mignone, A., JCP, **225**, 1427 (2007)
35. Mignone, A., Bodo, G., Massaglia, S., Matsakos, T., Tesileanu, O., Zanni, C., & Ferrari, A., ApJS, **170**, 228 (2007)
36. Miyoshi, T., & Kusano, K., JCP, **208**, 315 (2005)
37. Powell, K.G., ICASE report 94-24, 1994
38. Powell, K.G., Roe, P.L., Linde, T.J., Gombosi, T.I., & De Zeeuw D.L., JCP, **154**, 284 (1999)
39. Quirk, J.J., IJNMJ, **18**, 555 (1994)
40. Roe, P.L., JCP, **43**, 357 (1981)
41. Roe, P.L., & Balsara, D.S., SIAM JAM, **56**, 57 (1996)
42. Rusanov, V.V., J. Comput. Math. Phys. USSR, **1**, 267 (1961)
43. Ryu, D., & Jones, T.W., ApJ, **442**, 228 (1995)
44. Ryu, D., Miniati D., Jones, T.W., & Frank, A., ApJ, **509**, 244 (1998)
45. Tanaka, T., JCP, **111**, 381 (1994)
46. Toro, E. F., Spruce, M., & Speares, W., ShWav, **4**, 25 (1994)

47. Toro, E.F., Riemann solvers and numerical methods for fluid dynamics. Springer-Verlag, Berlin, (1997)
48. Torrilhon, M., JPlPh, **69**, 253 (2003)
49. Tóth, G., JCP, **161**, 605 (2000)
50. van der Vorst, H.A., SIAM JSSCom, **13**, 631 (1992)
51. Zachary, A.L., Malagoli, A., & Colella, P., SIAM JSCom, **15**, 263 (1994)

Part II

Hydrodynamic and Magneto-Hydrodynamic Instabilities

The Kelvin–Helmholtz Instability

E. Trussoni

INAF, Osservatorio Astronomico di Torino, 10025 Pino Torinese, Italy,
trussoni@oato.inaf.it

Abstract Since the discovery of jets in radio galaxies in early '70s, the Kelvin-Helmholtz Instability (KHI) has been envisaged as one of the fundamental processes that rule the dynamics of collimated outflows in various astronomical scenarios (AGN, Young Stellar Objects, compact galactic bodies, etc.). We outline here the main physical properties of the KHI following linear and non linear treatments, and considering typical conditions in astrophysical contexts: 2-D and 3-D geometries, magnetic fields, relativistic regimes. The formation of shocks and the turbulent mixing are the most important effects of the onset of the KHI that can affect the jets propagation and dynamics on galactic and extragalactic scales.

Keywords Magneto-hydrodynamics · instabilities · ISM: jets and outflows

Trussoni, E.: *The Kelvin–Helmholtz Instability.* Lect. Notes Phys. **754**, 105–130 (2008)
DOI 10.1007/978-3-540-76967-5_3   © Springer-Verlag Berlin Heidelberg 2008

1 Introduction

The concept of instability is fundamental to the understanding of several physical phenomena. As a very general definition, a system is unstable whenever a reaction to an external perturbation definitely modifies its structure through energy transfer. In an astrophysical context, this may occur in very different scenarios from the Earth's ionosphere to the farthest extragalactic objects. For example, solar flares or strong luminosity variabilities in compact objects (e.g., pulsars and active galactic nuclei) can originate from the release, through instabilities, of magnetic energy into particle acceleration. Moreover, the formation of stars or galaxies may be related to the onset of gravitational and radiative instabilities.

The Kelvin–Helmholtz Instability (KHI), that occurs at the interface between two fluids in relative motion, is one of the most important such processes. The main properties of the KHI are quite well known since ≈ 150 years, but its application to astrophysical phenomena until ≈ 30 years ago was quite limited. The KHI became a major argument of investigation for the interpretation of the newly discovered extragalactic radio sources that appeared as extended emitting structures symmetrically placed at tens of Kpc from a galactic nucleus. It was shown that the energetic requirements of such objects could be fulfilled assuming collimated jets connected the nucleus with these blobs. The key question was whether these beams could reach the radio lobes propagating through the environment safely against the onset of the KHI. The discovery in the following years of collimated outflows in other various environments, from galactic objects (young stars, pulsars, etc.) down to cometary tails interacting with the solar wind, attracted a lot of interest on the problems concerning the origin, dynamics, and evolution of such structures. As an obvious consequence, the KHI became one of the most relevant fields of activity in modern theoretical astrophysical research.

2 Definitions

Whenever a system is perturbed by an external disturbance, it can react by: (i) oscillating around the initial configuration indefinitely; (ii) oscillating with decreasing or (iii) increasing amplitude; (iv) monotonically moving away from the initial configuration. In the former two cases, the system is *stable*, in the (iii) is *overstable*, and in the (iv) it is *unstable*. It is clear that the configuration evolves modifying definitely its structure in the last two cases (for which the unique term *instability* is practically always used).

From a more physical point of view, we can investigate the stability of a configuration through energetic arguments. Considering its total energy, $E_{\mathrm{tot}} = E_{\mathrm{pot}} + E_{\mathrm{kin}} + \cdots = $ constant, a system is stable/unstable if a perturbation increases/lowers its potential energy E_{pot}.

Restricting to the typical conditions of astrophysical plasmas, we can define in general two main kinds of unstable processes:

Microinstabilities (kinetic). They are strictly correlated with the distribution function of the particles (e.g., non-Maxwellian). They are typical for space physics (e.g., ionosphere), heating processes, and interaction between particles and electromagnetic fields. They involve small scale lengths, with fast evolution with respect to the dynamical time scale of the system.

Macroinstabilities (fluid). They occur in configurations with an equilibrium distribution function for the particles, for which the fluid approximation holds. They are relevant whenever large scale phenomena are involved, with long evolutive times comparable or slower than the dynamical time scale.

The KHI belongs to this last kind of instability; then we can analyze its basic properties in the framework of a hydrodynamical and MHD treatment.

3 Equations and Mathematical Approach

The conservation of mass, magnetic flux, momentum, and energy, plus the Ampère law provide the system of equations governing the dynamics of a magnetized, ideal, and nonrelativistic flow with zeroth resistivity (ideal MHD approximation):

$$\frac{\partial \rho}{\partial t} + \nabla \cdot (\rho \mathbf{v}) = \nabla \cdot \mathbf{B} = \frac{\partial \mathbf{B}}{\partial t} - \nabla \times (\mathbf{v} \times \mathbf{B}) = 0 \tag{1}$$

$$\rho \left(\frac{\partial}{\partial t} + \mathbf{v} \cdot \nabla \right) \mathbf{v} = -\nabla p + \frac{1}{4\pi} (\nabla \times \mathbf{B}) \times \mathbf{B} - \rho \nabla \Psi \tag{2}$$

$$\left(\frac{\partial}{\partial t} + \mathbf{v} \cdot \nabla \right) P - \Gamma \frac{P}{\rho} \left(\frac{\partial}{\partial t} + \mathbf{v} \cdot \nabla \right) \rho = (\Gamma - 1)\mathcal{Q} \tag{3}$$

where Γ is the ratio of specific heats, Ψ the gravitational potential, and $\mathcal{Q}$ the volumetric rate of injection-minus-losses of energy in the plasma.

The standard method to treat such complex system (1), (2) and (3) is to study the evolution of small amplitude disturbances that perturb an equilibrium configuration (linear analysis). This is performed by expanding the physical variables Q_i as:

$$Q_i = Q_{0,i} + q_i, \quad q_i \ll Q_{0,i} \tag{4}$$

where $Q_{0,i}$ are the zeroth order, equilibrium values. Substituting the above relations in the original equations and neglecting all the terms $\mathcal{O}(q_i^2)$, we obtain a linear system for q_i, with coefficients functions of $Q_{0,i}$.

To solve this linearized system, we can refer to the energy criterion quoted in the previous section, looking for solutions that decrease the potential energy of the system: if they exist the configuration is unstable. This approach is suitable for complex structures; however, it does not provide information on the temporal growth rates of the instability. Furthermore, it does not work whenever velocities or dissipative processes are present.

The alternative approach, widely adopted for astrophysical plasmas, is more classical from the mathematical point of view: the perturbations are assumed as eigenfunctions of the linearized equations (*normal modes*) that must be solved fulfilling suitable boundary conditions. If the system is homogeneous, the eigenfunctions are the components of the Fourier expansion:

$$q_i = \frac{1}{2\pi} \int q_i^*(\mathbf{k}, \omega_{\mathbf{k}}) \exp[i(\mathbf{k} \cdot \mathbf{r} - \omega_{\mathbf{k}} t)] \mathrm{d}\mathbf{k} \, \mathrm{d}\omega_{\mathbf{k}} \tag{5}$$

and the system is solved for each of the components $\mathbf{k}, \omega_{\mathbf{k}}$. Considering that $\partial/\partial t = -i\omega$ and $\partial/\partial x_{1,2,3} = ik_{x_{1,2,3}}$ ($x_{1,2,3}$ are generic spatial coordinates) from the linearized equations, we obtain the *dispersion relation* (DR):

$$\mathcal{D}(\mathbf{k}, \omega_{\mathbf{k}}; Q_{0,i}) = 0 \tag{6}$$

Assuming $\mathbf{k}$ is real and $\omega = \mathrm{Re}\,\omega + i\mathrm{Im}\,\omega$, the above DR provides $\omega(\mathbf{k})$; for a fixed $\mathbf{k}$ the system is unstable if $\mathrm{Im}\,\omega > 0$. Considering the exponential increase of perturbation amplitudes, $\mathrm{Im}\,\omega$ is the growth rate and $t_i \approx 2\pi/\mathrm{Im}\,\omega$ is the time scale of evolution of the instability.

If the system is inhomogeneous in one direction (e.g., x_1), the perturbations can be expressed as oscillating functions only in time and in the remaining spatial components $x_{2,3}$, namely $q_i \propto f(x_1)\exp[i(k_{x_2}x_2 + k_{x_3}x_3 - \omega_{\mathbf{k}}t)]$. This means that a differential equation for $f(x_1)$ must be solved and that $\omega(\mathbf{k})$ cannot be provided by a simple expression of the DR like (6).

Few main points and warnings about the linear analysis must be outlined:

1. We could assume eigenvalue $\mathbf{k}$ (complex) as function of ω (real), obtaining the scale length of increase of instability: $l_i \approx 2\pi/|\mathbf{k}|_i$ (spatial analysis). Usually a temporal analysis is adopted, as we will do here.

2. The normal modes of analysis may not provide all the solutions. In particular, even if $\mathrm{Im}\,\omega = 0$,'transient' perturbations may be present that can be found through a different mathematical treatment (e.g., using Laplace transforms). Such solutions disappear on long times, however, could modify the equilibrium before damping.

3. When the instability is present, this does not necessarily means that the configuration is destroyed. In fact its growth rate can be so small (i.e., the instability evolves very slowly with respect to the dynamical time) to leave basically the system unaffected, which may be modified by other physical processes. Furthermore, the instability can saturate leading the system to a completely different but stable configuration. In more general, the linear approach cannot provide information on the global evolution of the instability; this can be obtained only by the complete solutions of (1), (2) and (3). This can be performed through numerical simulations using hydrodynamical codes that require advanced computing facilities. A lot of effort have been made in the last two decades to develop more and more sophisticated algorithms suitable for analyzing the global properties of supersonic flows.

In the following, we will discuss mainly the properties of the KHI in the linear regime with a few examples of nonlinear evolution for simple cases.

4 KHI: Linear Analysis

The fundamental features of the KHI can be outlined for an adiabatic and unmagnetized flow, without any gravitational field ($Q = \mathbf{B} = \Psi = 0$). In such a case, the system (1), (2) and (3) reduces to:

$$\frac{\partial \rho}{\partial t} + \nabla \cdot (\rho \mathbf{v}) = 0 \tag{7}$$

$$\rho \frac{D\mathbf{v}}{Dt} = -\nabla P, \quad \frac{DP}{Dt} = -\Gamma \frac{P}{\rho} \frac{D\rho}{Dt} \tag{8}$$

with $D/Dt = \partial/\partial t + (\mathbf{v} \cdot \nabla)$.

As equilibrium we will consider planar and cylindrical configurations. The former one allows quite a simple but standard mathematical treatment of the equations, while the second is more suitable for applications to astrophysical collimated outflows.

4.1 Planar Vortex Sheet

In a Cartesian frame of reference x, y, z, we adopt a planar discontinuity in the z, y plane (*vortex sheet*) separating two fluids in relative motion. The fluid 1, in the region $x > 0$, moves at constant velocity $\mathbf{v}_1 = u_0 \, \mathbf{z}$, while the fluid 2, in the region $x < 0$, is at rest ($\mathbf{v}_2 = 0$; see Fig. 1). We displace along $\mathbf{x}$ the vortex sheet by a small amount ξ perturbing all the physical quantities, see (4) and (5):

$$P = P_0 + p, \quad \rho = \rho_0 + \mu, \quad \mathbf{v} = u_0 \mathbf{z} + \mathbf{u} \tag{9}$$

$$p, \; \mu, \; |\mathbf{u}| \propto g(x) e^{i(k_z z + k_y y - \omega t)} \tag{10}$$

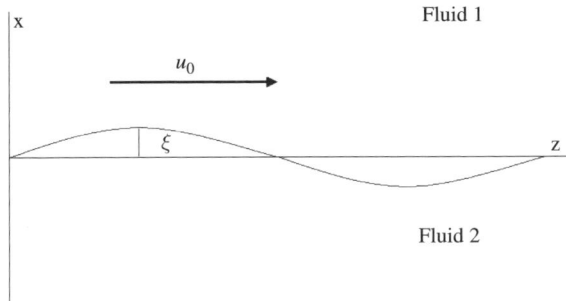

Fig. 1. Equilibrium configuration of a planar vortex sheet ($x = 0$) separating two fluids moving with velocity u_0 along $\mathbf{z}$ for $x > 0$ and at rest for $x < 0$, respectively. The vortex sheet is perturbed by a small displacement ξ along $\mathbf{x}$

with $k_{z,y}$ real and ω complex. If the system was homogeneous, it would be $g(x) = e^{ik_x x}$ and from (7) and (8) we would simply obtain the DR for the acoustic waves: $\omega = k_z s$ ($s = \sqrt{\Gamma P_0/\rho_0}$ is the sound speed). In the present case, conversely, we must solve the equations separately in the two fluids. For simplicity, the same density is assumed across the vortex sheet and we consider only perturbations propagating parallel to the velocity ($k_y = 0$).

Fluid 1. Equations (7) and (8) reduce to:

$$i\Omega\mu_1 = -\rho_0 \nabla \cdot \mathbf{u}_1, \quad i\Omega\rho_0\mathbf{u}_1 = -\nabla p_1, \quad p_1 = s^2\mu_1 \tag{11}$$

where $\Omega = k_z u_0 - \omega$. From the 1st and 3rd of (11) we get:

$$\nabla \cdot \mathbf{u}_1 = -i\frac{\Omega}{\rho_0 s^2}p_1 \tag{12}$$

while taking the divergence of the 2nd equation and considering that $\nabla^2 = \mathrm{d}^2/\mathrm{d}^2 x - k_z^2$, we end up with:

$$\frac{\mathrm{d}^2 p_1}{\mathrm{d}x^2} = \left(k_z^2 - \frac{\Omega^2}{s^2}\right)p_1 \tag{13}$$

which has solutions:

$$p_1 = A_1 e^{q_1 x}, \quad q_1 = \pm\left(k_z^2 - \frac{\Omega^2}{s^2}\right)^{1/2} \tag{14}$$

where A_1 and the sign of q_1 are still undetermined.

Fluid 2. Assuming $u_0 = 0$ and $\Omega \to -\omega$, and following the same steps as before we obtain the solutions:

$$p_2 = A_2 e^{q_2 x}, \quad q_2 = \pm\left(k_z^2 - \frac{\omega^2}{s^2}\right)^{1/2} \tag{15}$$

The constants A_1 and A_2 can be eliminated considering the $\mathbf{x}$ components of the momentum equation in the two fluids, the 2nd of (11):

$$i\Omega\rho_0 u_{x1} = -q_1 p_1, \quad i\omega\rho_0 u_{x2} = q_2 p_2 \tag{16}$$

and assuming pressure equilibrium across the vortex sheet ($x = 0$):

$$\frac{\Omega u_{x1}}{\omega u_{x2}} = -\frac{q_1}{q_2} \tag{17}$$

Through the perturbed displacement of the vortex sheet $\xi(z,t)\mathbf{x}$ (common for the two fluids), we can now eliminate $u_{x1,2}$ considering that $u_{x1} = i\Omega\xi$ and $u_{x2} = -i\omega\xi$, and obtain the DR:

$$\mathcal{D}(k_z, \omega; u_0; s^2) \equiv \Omega^2 q_2 - \omega^2 q_1 = 0 \tag{18}$$

The signs of q_1 and q_2 depend on the solution itself and on the boundary conditions, and we have two possible choices:

(a) If ω is real the system is stable: the sign must be selected such that the perturbations are outgoing traveling waves to avoid a source of energy at infinity (Sommerfield conditions).

(b) If ω is complex or imaginary the system is unstable: then the sign must be consistent with the amplitude of the perturbations vanishing for $x \to \pm\infty$.

Solutions of the DR

To solve the DR, it is convenient to deal with nondimensional variables; accordingly we define:

$$\Phi \equiv \text{Re}\,\Phi + i\text{Im}\,\Phi = \frac{\omega}{k_z s}, \quad M = \frac{u_0}{s} \tag{19}$$

where M is the Mach number. The time scale of the instability is then given by $t_i = 2\pi/k_z s\,\text{Im}\,\Phi$.

By squaring (18), the DR reduces to a 6th polynomial that can be easily solved. However in this way spurious roots are introduced and it must be verified that each solution fulfills the original DR (18). It turns out that of all the 6 roots of the polynomial just these are acceptable:

$$\Phi = \frac{M}{2} \pm i \left[(M^2 + 1)^{1/2} - \left(\frac{M^2}{4} + 1 \right) \right]^{1/2} \tag{20}$$

$$\Phi = \frac{M}{2}, \quad M > 2 \tag{21}$$

The only unstable solution is (20) with the sign '+', and its main properties, as function of the Mach number, are the following (see Fig. 2):

M ≪ 1. For largely subsonic velocities, corresponding to the incompressible regime, (20) reduces simply to:

$$\Phi = \frac{M}{2}(1 \pm i) \tag{22}$$

which means that the vortex sheet is always unstable against the KHI, and the growth rate linearly increases with the velocity. We notice further that the phase velocity of the perturbations ($= u_0/2$) is just the Doppler shift; then in a system with the two fluids moving at relative velocities $\pm u_0/2$ the amplitude of the modes is monotonically increasing (these are true unstable, and not overstable modes; see Sect. 2).

M ≳ 1. Also a transonic flow is always unstable but the increase of $\text{Im}\,\Phi$ with M is slower; it reaches a maximum at $M \approx 1.7$ and then decreases.

M ≫ 1. For $M > \sqrt{8}$ the two roots of (20) become real ($\text{Im}\,\Phi = 0$) with the perturbations transforming into two traveling sound waves: highly supersonic flows are always stable. However, it is possible to see that the reflection and transmission coefficients of these waves across the vortex sheet can diverge in particular conditions. This means that these perturbations are marginally stable.

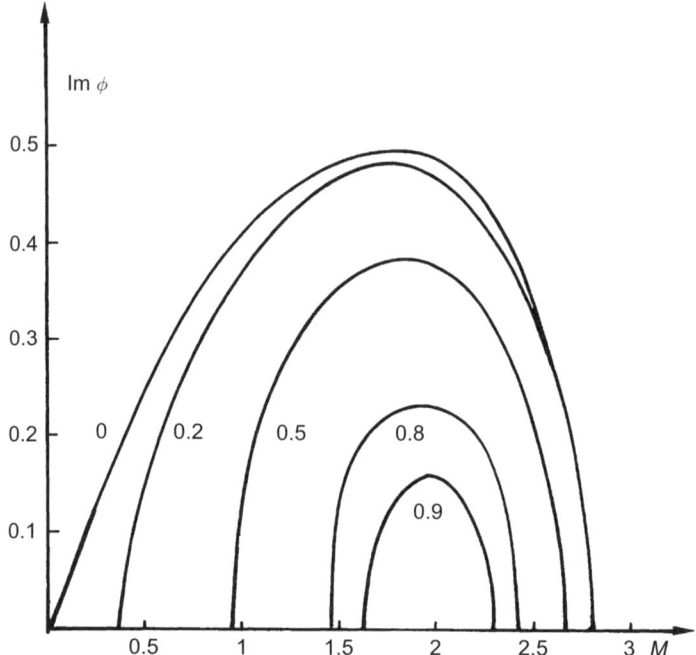

Fig. 2. Plot of the nondimensional growth rate Im Φ vs M of a perturbation propagating along the fluid velocity for a vortex sheet configuration, and for unmagnetized and magnetized flows (labels are different values of v_A/s)

Transverse Propagation, Fluid Densities, and Magnetic Field

We shortly summarize how the KHI properties are modified when some of the above simplifying assumptions are released.

If the modes do not propagate along the velocity direction ($k_y \neq 0$), the previous mathematical results are nonmodified provided that $M \to M_{\mathrm{eq}} = Mk_z/\sqrt{k_z^2 + k_y^2}$ and $\Phi \to \omega/\left(s\sqrt{k_z^2 + k_y^2}\right)$. From these new definitions, it appears evident that the onset of the KHI does not depend on the full velocity but depend on its component along the direction of propagation of the perturbation. By increasing k_y, we move towards the incompressible regime (for $k_y \gg k_z$ is $M_{\mathrm{eq}} \ll 1$): highly supersonic flows may be unstable against oblique disturbances, but the growth rate vanishes for $k_z \to 0$, and modes propagating parallel to $\mathbf{u}_0$ are stable as they do not feel any velocity jump.

The general features of the KHI also do not change if there is a contrast of density across the vortex sheet. The growth rate of the instability decreases by a factor $\propto \sqrt{\rho_{02}/\rho_{01}}$ ($\rho_{02} < \rho_{01}$), and the stability cut off for highly supersonic flows is always present.

When the magnetic field is included, $\mathbf{B} = \mathbf{B}_0 + \mathbf{b} \neq 0$, the linearization process refers now to the full MHD system (1), (2), and (3) but the procedure to get the DR is the same as in the hydrodynamical case: now the DR is a polynomial of 10th degree. The effect of **B** on the development of the KHI strictly depends on its geometry. When the velocity and magnetic field are perpendicular the features of the instability are basically unmodified if we redefine the Mach number as $u_0/\sqrt{s^2 + v_A^2}$, where v_A is the Alfvèn velocity $(= B_0/\sqrt{4\pi\rho_0})$. Conversely, the magnetic field has a stabilizing effect when it is along the flow speed (see Fig. 2). For parallel propagation of modes ($k_y = 0$) and for $M \leq 2v_A/s$ it is Im $\Phi = 0$, i.e. subsonic magnetized flows are stable. Moreover, the supersonic cut off decreases: $M = \sqrt{8} \rightarrow 2$, and for $v_A/s \geq 1$ the two cut off coincide: highly magnetized flows are always stable against the KHI (for $k_y \neq 0$ hold the previous considerations on oblique mode propagation with $M \rightarrow M_{\mathrm{eq}}$).

4.2 Shear Layer

It is clear from the first term of (19) that the growth rate diverges for modes with vanishing wavelengths ($t_i \rightarrow 0$ for $k_z \rightarrow \infty$); this is related to the lack of any typical scale length in the equilibrium configuration. From the mathematical point of view, this means that the KHI is an 'ill posed' problem in the case of vortex sheet; this is removed if we introduce a continuous transition of velocity across the two fluids.

Now the mathematical approach is different: working out the linearized system with the same geometry as for the vortex sheet (velocity parallel to **z** and varying along **x**), we obtain a unique differential equation that governs the evolution of the perturbations in the whole system:

$$\frac{\mathrm{d}^2 p}{\mathrm{d}x^2} + g(x, M, M'; \Phi, k_z a)\frac{\mathrm{d}p}{\mathrm{d}x} + f(x, M, M'; \Phi, k_z a)p = 0 \qquad (23)$$

where g and f are general functions (the unmagnetized case has been considered) and a is the scale length of variation of the Mach number ($a = M/M'$, $M' = \mathrm{d}M/\mathrm{d}x$). As we are dealing with an ideal flow, the shape of $M(x)$ is a free function; in general the transition is assumed from purely linear, $M(x) \propto (x - x_0)/a$, to smoother profiles, $M(x) \propto \tanh[(x - x_0)/a]$. With this assumption (23) is integrated searching those values of the eigenvalue Φ that allow to fulfill the boundary conditions for the eigenfunction, i.e. $p \rightarrow 0$ for $x \rightarrow \pm\infty$.

The properties of the solutions depend on the interplay between M and $k_z a$. In subsonic and transonic flows, $M \lesssim 1-1.5$, the large wavelength modes ($k_z a \ll 1$) are basically unaffected by the shear layer with respect to the vortex sheet configuration. Conversely, perturbations with wavelengths comparable to or shorter than the transition scale length ($k_z a \gtrsim 1$) are stabilized, almost independently on the velocity profile (Fig. 3, left panel).

The behavior of the KHI is more complex for supersonic speeds. Modes with $k_z a \gtrsim 1$ tend always to be damped while the stability cut off found

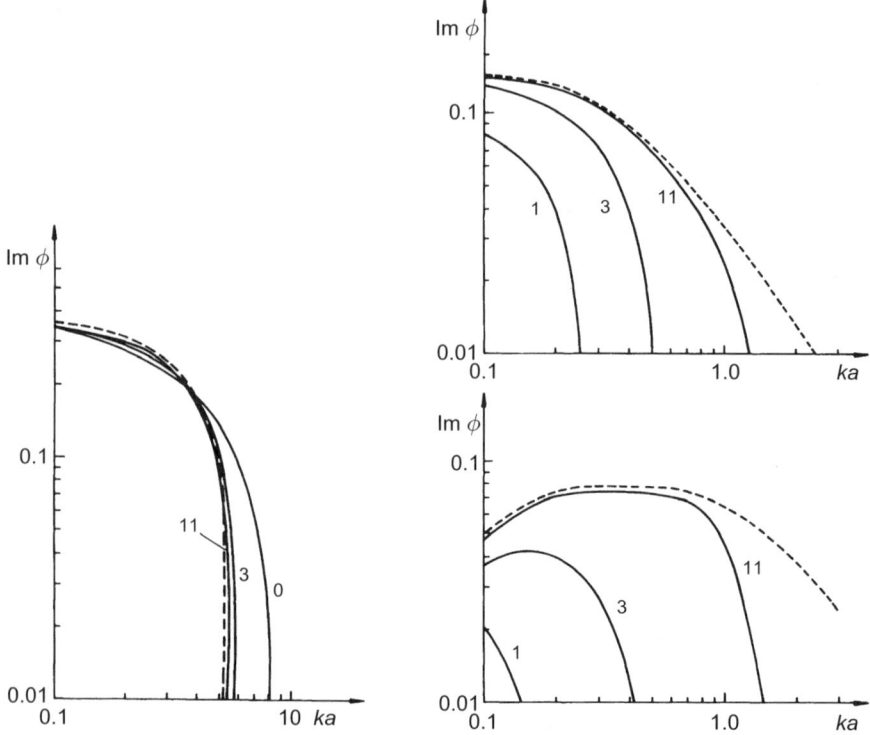

Fig. 3. Planar shear layer. Plot of Im Φ vs $k_z a$ for three values of the Mach number and different transition profiles from linear (*dotted*) to more and more smoothed for decreasing labels (label '1' refers to the 'tanh' profile). *Left panel*: Stabilizing effect of the shear layer in a transonic flow ($M = 1$) by increasing $k_z a$. *Right panels*: Stabilizing effect in supersonic flows (*upper panel*, $M = 3$; *lower panel*, $M = 5$) of the wavenumber $k_z a$ on the destabilized modes which are overstable in the limit of the supersonic vortex sheet ($M > \sqrt{8}$)

for $M > \sqrt{8}$ disappears (Fig. 3, right panels). In fact the behavior with the Mach number is quite peculiar: for $M > 2$ the solution given by (21) becomes unstable, and for $M \approx \sqrt{8}$ it merges with the overstable modes that are now destabilized by the shear (it turns out from the phase velocity, Re Φ, that these are traveling, i.e., overstable perturbations). Then highly supersonic flows can always be unstable (even though with much lower growth rates than in the vortex sheet case), but this strictly depends on the shape of the velocity transition. For linear profiles, a highly supersonic shear layer is never stabilized, however, large are M and $k_z a$. Conversely, for smoother transitions, an upper cut off value of $k_z a$ is found for the instability, but always for $M > \sqrt{8}$. In the opposite limit of vanishing layer, $k_z a \to 0$, we converge to the vortex sheet configuration: Im $\Phi \to 0$ and these modes degenerate into the overstable perturbations.

4.3 Cylindrical Geometry

The DR is obtained following an analogous method as in the planar case. The equilibrium configuration is a jet of infinite length and radius a, with axis and constant velocity along $\mathbf{z}$ (cylindrical coordinates r, ϕ, and z are adopted) and separated by the external unmoving environment by a discontinuity. The modes are chosen as:

$$p,\ \mu,\ |\mathbf{u}| \propto g(r)e^{i(kz+n\phi-\omega t)} \qquad (24)$$

where the azimuthal number n defines the symmetry of the perturbations. For $n = 0$, $n = 1$, and $n > 1$, we have the axisymmetric pinching, helical, and fluting modes, respectively (see Fig. 4).

From the linearized system (7) and (8), we now obtain two Bessel equations of order n for the perturbed pressures $p_{i,e}$ inside and outside the jet. These equations are solved considering suitable boundary conditions: p_i must be regular on the axis, while the amplitude of p_e must vanish for $r \to \infty$. Accordingly, for a perturbation of azimuthal number n, the solutions for $r < a$ and $r > a$ are given by the Bessel, J_n, and Hankel, $H_n^{(1)}$, functions, respectively. Matching the two solutions on the jet boundary ($r = a$), we obtain the following DR:

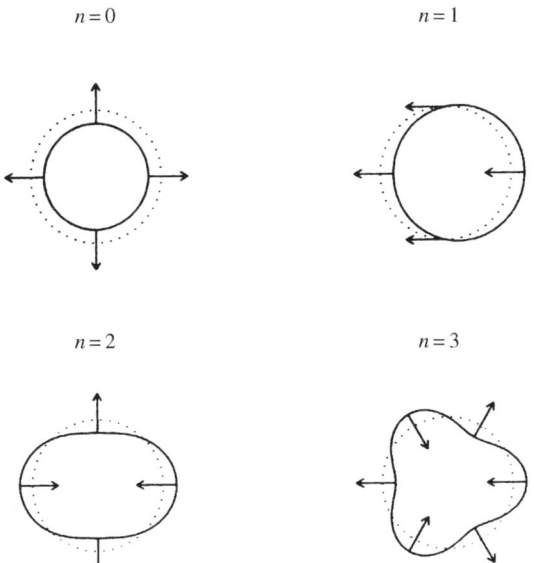

Fig. 4. Projection of the cross sectional area of a cylinder showing different kinds of perturbations: pinching ($n = 0$), helical ($n = 1$), and fluting ($n > 1$)

$$\nu(\Phi - M)^2 \Delta_e \frac{H_n'^{(1)}(ka\Delta_e)}{H_n^{(1)}(ka\Delta_e)} - \Phi^2 \Delta_i \frac{J_n'(ka\Delta_i)}{J_n(ka\Delta_i)} = 0 \qquad (25)$$

$$\Delta_i^2 = (\Phi - M)^2 - 1, \quad \Delta_e^2 = (\Phi^2 - \nu)/\nu \qquad (26)$$

where primes indicate derivatives with respect to the argument and ν the ratio of the beam to external densities (for $\nu > 1$ or < 1, we have a 'dense' or 'light' jet, respectively).

Considering the asymptotic expansions of the Bessel and Hankel functions, we can have expressions of the DR much simpler than (25). For very long wavelengths, $ka\Delta_{i,e}/n \ll 1$, the DR for helical and fluting modes ($n > 0$) reduces to:

$$\nu(\Phi - M)^2 = -\Phi^2 \quad \rightarrow \quad \Phi = \nu M \frac{1 \pm i/\sqrt{\nu}}{1 + \nu} \qquad (27)$$

which is the same expression found for a subsonic planar vortex sheet across equal fluids [$\nu = 1$, see (22)]. This relation implies that these axisymmetric perturbations are always unstable without any stability cut off for supersonic speeds. The growth rate for very light ($\nu \ll 1$) or very dense jets ($\nu \gg 1$) decreases as $\nu^{1/2}$ or $\nu^{-1/2}$, respectively: as in the planar vortex sheet the fastest growing KHI occurs in beams moving through environments with similar densities.

In the opposite case of very short wavelengths, $ka\Delta_{i,e}/n \gg 1$, (25) becomes:

$$\exp\left[2i\left(ka\Delta_i - \frac{2n - N + 1}{4}\pi\right)\right] = \frac{1 + F_1(\Phi)}{1 + F_2(\Phi)} \qquad (28)$$

where F_1 and F_2 are regular analytic functions and $N \geq 0$. It is clear from the form of (28) that various solutions are possible for the same set of parameters: a completely different behavior with respect to the planar vortex sheet.

The numerical solution of (25) confirms these asymptotic properties, as shown in Figs. 5 and 6, where we plot $\text{Im}\,\Phi$ against M and ka for pinching and helical perturbations, respectively (these patterns are not strongly modified for $n > 1$). It turns out that two main kinds of unstable perturbations are present in a cylindrical beam. The so-called *ordinary*, or *surface* modes are the equivalent of the unstable mode of a planar vortex sheet, with the amplitude exponentially decreasing moving away from the jet surface. However, the geometry introduces two main differences: there is any stability cut off for supersonic speeds, whichever is the value of ka and M; and for small ka and M we find $\text{Im}\,\Phi(n = 0) \ll \text{Im}\,\Phi(n > 0)$. For long wavelength asymmetric modes, the growth rate becomes the one given by (27).

For $M > 2$ and not extremely long wavelengths, a new series of modes are found mixed with the ordinary perturbation and with growth rates basically

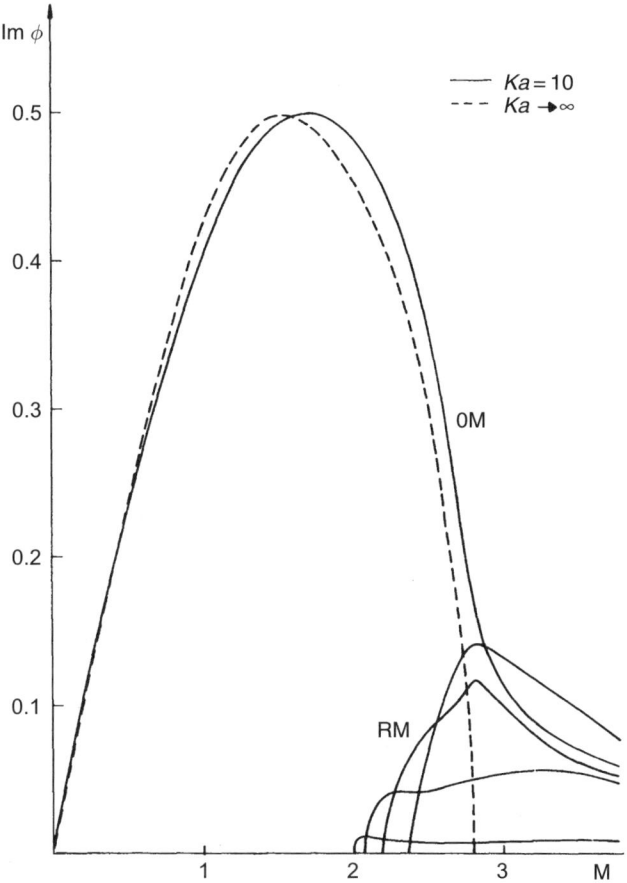

Fig. 5. Plot of Im Φ vs M for pinching modes of short wavelength ($n = 0$, $ka = 10$). The dotted line represents the planar vortex sheet limit. With OM and RM, we label ordinary and reflected modes, respectively (only the first four RM are plotted)

independent of n [the DR for these kind of perturbations with short wavelengths is given by (28)]. As for the new mode found in a supersonic planar shear (Sect. 4.2), this instability is related to the overstable perturbations found in the limit of a vortex sheet with $M > \sqrt{8}$. These disturbances in some conditions can have multiple reflections on the cylinder wall leading to instability. The properties of these modes (called *reflected*) are strictly related to the wavenumber and the Mach number, and the main consequence is that collimated outflows are always unstable (see Figs. 5 and 6).

If a magnetic field is included, its effect depends on the geometry of **B**: as a reasonable configuration it is generally assumed a longitudinal component inside ($B_{0,z} \neq 0$ for $r < a$) and a toroidal component outside the jet ($B_{0,\phi} \neq 0$ for $r > a$). As in the planar case, a magnetic field along the beam axis

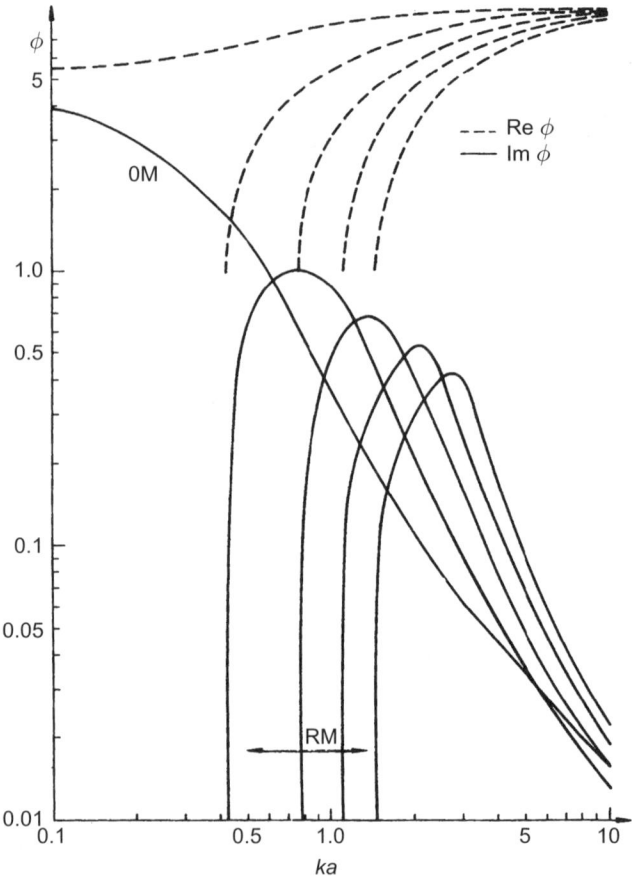

Fig. 6. Plot of $\mathrm{Im}\,\Phi$ and $\mathrm{Re}\,\Phi$ vs ka for helical modes in a highly supersonic jet ($n = 1$, $M = 10$). With OM and RM, we label ordinary and reflected modes, respectively (only the first four RM are plotted)

stabilizes the KHI, mainly in slow jets and for long wavelength ordinary modes. An azimuthal magnetic field has conversely an opposite trend enhancing the instability, and this is expected as a typical property of magnetized cylindrical equilibria. It is well that for nonmoving MHD cylindrical configurations ($u_0 = 0$) $B_{0,\phi}$ triggers the onset of the well known sausage ($n = 0$) and kink ($n = 1$) unstable modes that can be stabilized by a strong enough longitudinal magnetic field.

4.4 KHI: Physical Properties

We have outlined the general properties of the KHI, mainly from the mathematical point of view; now we address to the simple question: what is the physical mechanism that drives the instability process?

Lets consider the perturbed planar vortex sheet separating the two flows as shown in Fig. 1. When we slightly displace this surface by ξ, a pressure unbalance is created across the boundary due to the centrifugal force acting on the flow moving along the corrugated sheet. The strength of this unbalance grows with the perturbation amplitude as $\xi \propto dp/dx$, then once perturbed, the system moves indefinitely away from the initial configuration. If a shear layer is introduced between the two fluids, the coherence between the amplitude and the pressure gradient is lost for perturbations with wavelength comparable or smaller than the scale length of the velocity variation. The modes with $ka \gtrsim 1$ are then stabilized. It is also clear the stabilizing effect of a magnetic field parallel to the flow direction: the fieldlines act as rigid 'wires' that counteract the deformation of the discontinuity.

The physical origin of unstable, reflected perturbations, sometimes also called 'negative energy' modes is completely different. We have seen that in supersonic jets overstable modes in some conditions can have multiple reflections on the jet boundary. There, they can resonantly couple with external acoustic waves that travel away from the jet surface subtracting energy from the internal beam that reacts to this loss of energy becoming unstable. A similar resonant process occurs in a planar shear layer. The different nature of the reflected modes is clear: they are not perturbation localized near the velocity transition (as ordinary modes), but true acoustic waves propagate inside and far outside the beam. If a magnetic field is considered, it is possible to see that reflected modes are associated with fast magnetosonic waves (also slow magnetosonic waves can be unstable for some values of the parameters, but they have always very low growth rates).

All previous considerations hold in the framework of an ideal, classical and, adiabatic MHD treatment, but more general physical scenarios could be assumed in an astrophysical context. In such a case, new kinds of instabilities can arise that mix and interact with the KHI, as in the case of flute and sausage instabilities for a magnetized, nonmoving cylinder that we quoted at the end of Sect. 4.3. If in (2) $\nabla\Psi \neq 0$ the configuration may undergo Rayleigh–Taylor and Jeans instabilities, while heating and radiative losses, $Q \neq 0$ in (3), lead to the onset of radiative instabilities.

We discuss in more detail the properties of the KHI in relativistic regime, that is typical in several astrophysical phenomena.

Relativistic Flows

The conservation laws in (2) and (3) in the special relativistic ideal hydrodynamics regime become:

$$\gamma \left(\frac{\partial}{\partial t} + \mathbf{v} \cdot \nabla \right) \rho + w \left[\frac{\partial \gamma}{\partial t} + \nabla \cdot (\gamma \mathbf{v}) \right] = 0 \qquad (29)$$

$$\gamma^2 w \left(\frac{\partial}{\partial t} + \mathbf{v} \cdot \nabla \right) \mathbf{v} = -\nabla P - \frac{\mathbf{v}}{c^2} \frac{\partial P}{\partial t}, \qquad P \rho_r^{-\Gamma} = \text{constant} \qquad (30)$$

where $\gamma = \sqrt{1 - \beta^2}$ ($\beta = |\mathbf{v}|/c$) is the bulk Lorentz factor, $w = \rho + p/c^2$ the enthalpy, $\rho_r = m_0 n$ the rest mass density (with m_0 and n the particle rest mass and number density in the fluid rest frame, respectively), and $\rho = \rho_r(1 + e/c^2)$ is the relativistic density (with e the specific internal energy). The sound speed is defined as $s^2 = \Gamma P/w$, and the relation between P and e is given by $P = (\Gamma - 1)e\rho_r$. In plasmas with ultrarelativistic temperatures, $\rho \gg \rho_r$, the adiabatic relation modifies as $P \propto \rho^{5/3} \to \propto n^{4/3}$, which is just the second of (30) with $\Gamma = 4/3$; then for the sound speed we have $s = c/\sqrt{3}$ and the classical Mach number has the upper limit $M \leq \sqrt{3}$ (for mildly relativistic temperatures, the equation of state is very complicated).

Linearizing (29) and (30) following the same procedure as in previous Sects. 4.1 and 4.3, we obtain DR which are similar to (18) and (25) for planar and cylindrical geometries, respectively.

Referring to cylindrical beams, more interesting in an astrophysical context, (25) and (26) now become (Δ_e does not change):

$$\gamma^2 \nu (\Phi - M)^2 \Delta_e \frac{H_n'^{(1)}(ka\Delta_e)}{H_n^{(1)}(ka\Delta_e)} - \Phi^2 \Delta_i \frac{J_n'(ka\Delta_i)}{J_n(ka\Delta_i)} = 0 \tag{31}$$

$$\Delta_i^2 = \gamma^2 \left[(\Phi - M)^2 - \left(1 - \frac{\beta^2}{M}\Phi \right) \right], \quad \nu = w_i/w_e \tag{32}$$

Now the DR depends on the bulk Lorentz factor and the ratio of the inner and outer enthalpies (instead of the densities). Then we can have different 'relativistic regimes': (i) a cold flow ($\rho \to \rho_r$) moving at highly relativistic velocities ($\gamma \gg 1$), (ii) a hot beam ($\rho \gg \rho_r$) with low velocity ($\gamma \approx 1$), and (iii) a hot, highly relativistic jet ($\rho \gg \rho_r$, $\gamma \gg 1$).

We can easily obtain information on the relativistic effects, unstable long wavelength helical and fluting ordinary perturbations, in which case the DR is simply given by [see (27) for the classical case]:

$$\gamma^2 \nu (\Phi - M)^2 = -\Phi^2 \quad \to \quad \Phi = \gamma^2 \nu M \frac{1 \pm i/\gamma\sqrt{\nu}}{1 + \gamma^2 \nu} \tag{33}$$

In highly relativistic regime ($\gamma \gg 1$ and/or $\nu \gg 1$), $\mathrm{Im}\,\Phi \propto (\gamma\sqrt{\nu})^{-1}$, which means that these unstable perturbations are damped in hot or cold relativistic jets. Numerical solutions of (31) confirm this trend also for intermediate wavelengths, including pinching modes: when the jet velocity c and/or its temperature is very high, $\rho \gg \rho_r$, the larger inertia of the plasma contrasts the growth of the amplitude of perturbations in the transition boundary.

The behavior of the reflected modes is more complex, in which case the instability is related to the resonant coupling of acoustic waves at the jet surface. Now the wavelength of the perturbations inside the beam are affected by the relativistic velocity, and it turns out from the solutions of (31) that reflected modes with smaller wavenumbers are destabilized for the velocity approaching the speed of light. In particular, by increasing γ the pattern of the perturbations shown in Fig. 6 is shifted to smaller values of ka with almost unchanged values of $\mathrm{Im}\,\Phi$. The final consequence is a decrease of the physical growth rate, then also reflected modes are damped in ultrarelativistic jets.

5 KHI: Nonlinear Evolution

The linear analysis of the fluid and MHD equations allowed one to investigate the main physical conditions for the onset of the KHI in different contexts. However, a general view on the global evolution of the unstable system requires the direct treatment of the Eqs. (1), (2), and (3), and this is performed only through hydrodynamical and MHD simulations using advanced numerical codes. We present, in the following, the main properties of the nonlinear evolution of the KHI in the simplest cases of a planar layer and a cylindrical beam. The numerical results have been obtained using, for the simulations, a hydrodynamical code based on the Piecewise Parabolic Method algorithm (PPM; we do not discuss here the basic principles of different numerical codes, that can be found in other chapters of this book).

5.1 Shear Layer

In a discretized domain, a planar vortex sheet is approximated by a layer with thickness equal to the size of the smallest grid point. However, the KHI in this case is an 'ill posed' problem (see Sect. 4.2), then we expect that numerical noise with dimension comparable with these small grids would sharply increase leading to unphysical spurious solutions. To avoid this effect, we assume, for the initial equilibrium, a configuration similar to that discussed in Sect. 4.2: a homogeneous flow in pressure equilibrium with velocity along the **x** axis and with a 'tanh' profile in the **y** direction: $v_x(y) = v_0 \tanh[(y - y_0)/a]$. The total jump of velocity is $2v_0$ and a is the scale length of the transition shear. In this way, all the rapidly increasing, small size disturbances are damped.

This configuration is perturbed assuming a transversal sinusoidal velocity with small amplitude, exponentially decaying moving away from the shear layer:

$$u_y(x,y) = \epsilon \sin(2\pi x/\lambda)\, \mathrm{e}^{-[2\pi(y-y_0)/\lambda]}, \quad \epsilon \ll v_0 \qquad (34)$$

The spatial coordinates are given in units of half wavelength of the perturbation, and the temporal coordinate in units of the time elapsed to cross a unit length moving at the sound speed (dynamical time). The simulations have been performed for transonic ($M = 1$) and supersonic velocities ($M = 3$), assuming free boundary conditions far from the layer and periodic boundary conditions along the flow.

For transonic velocities the perturbation evolves as predicted by the linear theory for few dynamical times, then its amplitude stops its growth, and a new permanent equilibrium is attained. The final configuration has a *cat's eye* structure: a vortex shape with a minimum density inside and the pressure and centrifugal force balancing on the wall, with no energy outflow from the central region (Figs. 7 and 8, upper panels). The number of vortices depends on the wavelengths of the perturbation: for short wavelengths several vortices grow that merge forming a final unique steady structure. A spatial Fourier analysis shows that only one scale is present in the velocity field comparable with the size of the domain.

For supersonic velocities, the perturbations are overstable modes destabilized by the shear, then their evolution is completely different. We see in Figs. 7 and 8 (lower panels) that the central density oscillates with increasing amplitude while outgoing waves originate from the layer that, after some dynamical times, evolve into shock waves with increasing strength. This behavior implies that a spectrum of structures with different scale lengths are present in the system as deduced from the spatial Fourier analysis. The evolution is slower for transonic velocities as expected from the lower growth rate found from the linear analysis; however, no asymptotic configuration is attained and the energy is indefinitely carried away from the layer.

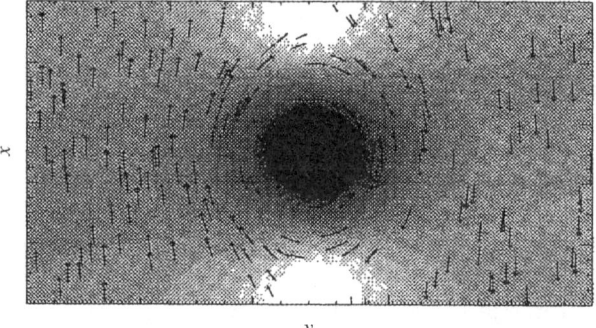

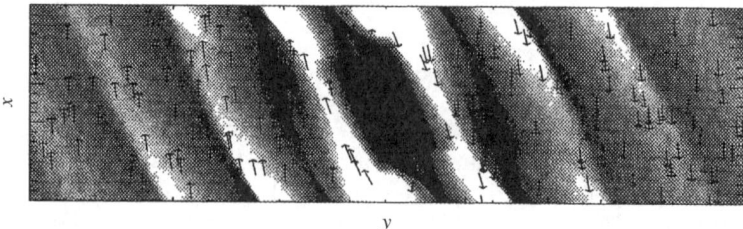

Fig. 7. Two-dimensional maps of the density distribution for a sheared 'tanh' flow with $M = 1$ at $t = 6.1$ (*upper panel*) and $M = 3$ at $t = 12.1$ (*lower panel*). Dark zones indicate low density regions; the arrows represent the velocity field, initially parallel to the **x** axis

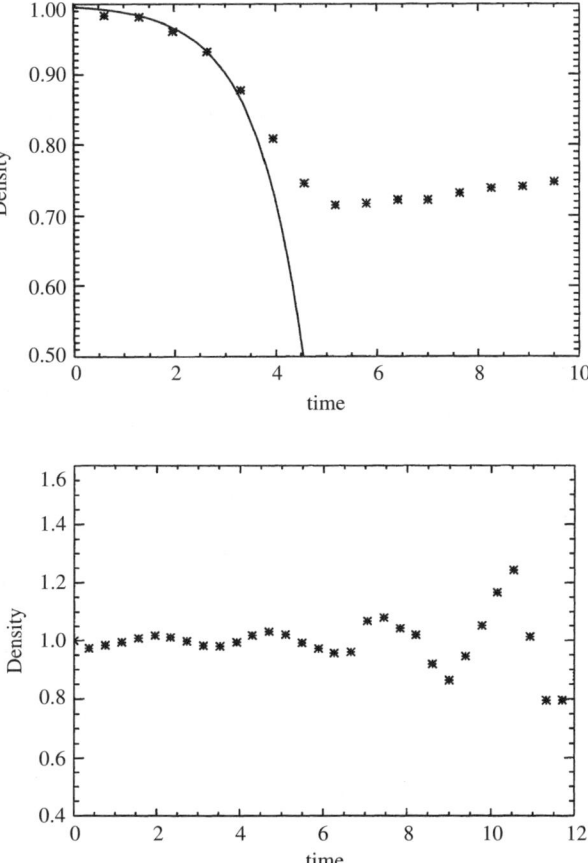

Fig. 8. Plot of t vs the density at the center of the shear layer for $M = 1$ (*upper panel*) and $M = 3$ (*lower panel*). The solid line for the transonic case represents the evolution expected from the linear theory

5.2 Cylindrical Jet

To simulate the evolution of a jet against the KHI, we assume the same equilibrium configuration as in Sect. 4.3, including a velocity and density profile across the beam and external medium given by:

$$v_z(r) = v_0/\cosh[(r/a)^m] \qquad \rho(r) = \nu - \nu/\cosh[(r/a)^m] \tag{35}$$

where m is related to the steepness of the transition of the variables across the sheared surface and ν is the density contrast between the center of the jet and outer environment: $\nu > 1$ or $\nu < 1$ for light or dense beams, respectively (this definition of ν is the opposite of that adopted in Sect. 4.3).

For the perturbations, we assume a more general form than in the case of planar shear. As we have seen, several unstable modes are present in highly supersonic jets. In this case, a discrete spectrum of perturbations is more suitable to simulate the nonlinear evolution of the KHI. Limiting our discussion to axisymmetric pinching disturbances ($n = 0$), it has been assumed for the transverse perturbed velocity:

$$u_r(r, z) = \epsilon/n_0 \sum_1^{n_0} \sin(nk_0 z)e^{-[(r-a)/\delta]^2}, \quad \epsilon \ll v_0 \tag{36}$$

The longest wavelength is equal to the longitudinal length D of the computational domain ($k_0 = 2\pi/D$) while the choice of smallest wavelength (n_0k_0) is limited by the grid resolution. Furthermore, it has been assumed that the amplitude of the modes exponentially decay on a scale δ transverse to the beam surface. We have adopted periodic boundary conditions on the left and right side of the domain ($z = 0$ and $z = D$), symmetric or anti symmetric boundary conditions on the jets axis ($r = 0$), and free conditions on the opposite boundary, far from the beam ($r \gg a$).

The evolution of the instability is discussed for a dense, highly supersonic jet ($M = 20$, $\nu = 0.3$) and a light, slower jet ($M = 5$, $\nu = 3$). For the parameters of the perturbations appearing in (36), we have assumed in both cases $m = 8$, $\delta = 0.2$ and $n_0 = 12$, i.e., 12 unstable modes are overimposed on the beam (see Fig. 9).

The simulations show that the KHI develops following these three main phases.

(1) *Linear.* The amplitudes of the modes increase as predicted by their growth rate obtained from the linear study: the prevailing scale corresponds to the perturbation with the largest growth rate (see Fig. 10).
(2) *Expansion.* The prevailing perturbation transforms into internal shocks: they heat the intra shock region that expands but remain well separated from the outer medium.
(3) *Mixing.* The shocks evolve leading to mixing with entrainment of the external medium inside and outwards diffusion of the jet gas: as a consequence the velocity shear widens and the density contrast between the jet and the environment decreases. The prevailing scale in this phase corresponds to the unstable mode with the longest wavelength (see Fig. 10).

The temporal intervals between these phases and the final configurations strongly depend on the jet parameters. For dense and highly supersonic jets the evolution is quite fast and full mixing with the external medium is attained at $t \approx 17$ (see Fig. 11). As a result the velocity shear widens but still maintains a coherent collimated pattern: a complete disruption of the jet may occur but on quite long time scales. In highly supersonic but light jets, the expansion is larger and they appear very well separated from the external medium for long times, with the mixing occurring much later than in dense beams. The opposite trend is found in slower jets (see Fig. 12): the surface is less inflated

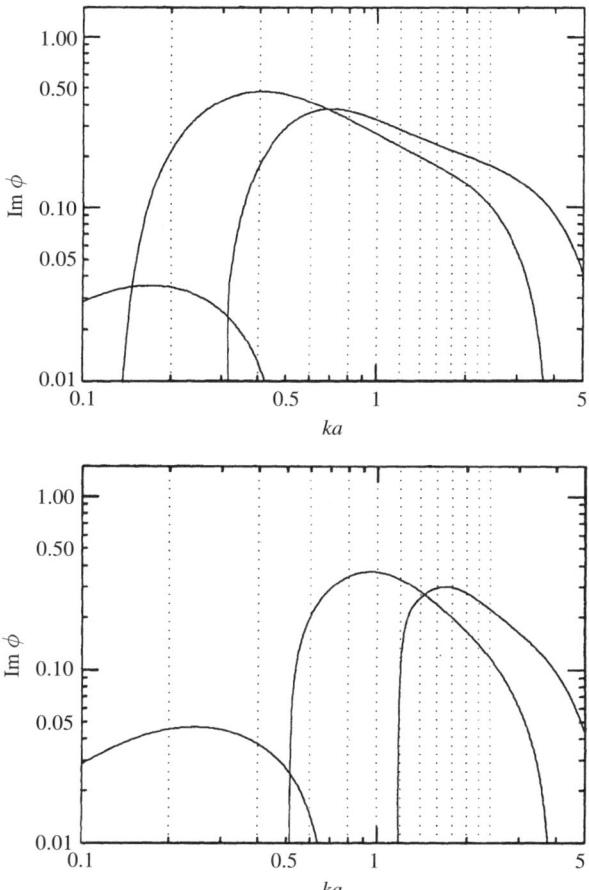

Fig. 9. Plot of $\mathrm{Im}\,\Phi$ vs ka for a pinching perturbations ($n = 0$), and $M = 20$, $\nu = 0.3$ (dense jet, *upper panel*), and $M = 5$, $\nu = 3$ (light jet, *lower panel*). Only the ordinary and the first two reflected modes are shown. The dotted vertical bars indicate the values of ka assumed in the initial perturbations (36)

but more deeply deformed (unstable modes have shorter wavelengths), with the consequent final and almost complete disruption of the beam. In particular, the mixing phase begins at $t = 19$, and for $t \geq 25$ strong vortices are formed and the flow is basically completely decelerated.

In conclusion, the KHI deeply affects the structure of supersonic cylindrical beams and has two main effects: the formation of transverse shocks, with consequent strong heating of the plasma, and the mixing of external and beams matter with the onset of peculiar structures and morphologies, and the (earlier or later) final disruption of the collimated beam. These basic properties qualitatively hold even if more complex configurations are considered, including 3-D geometry, magnetic field, and relativistic velocities.

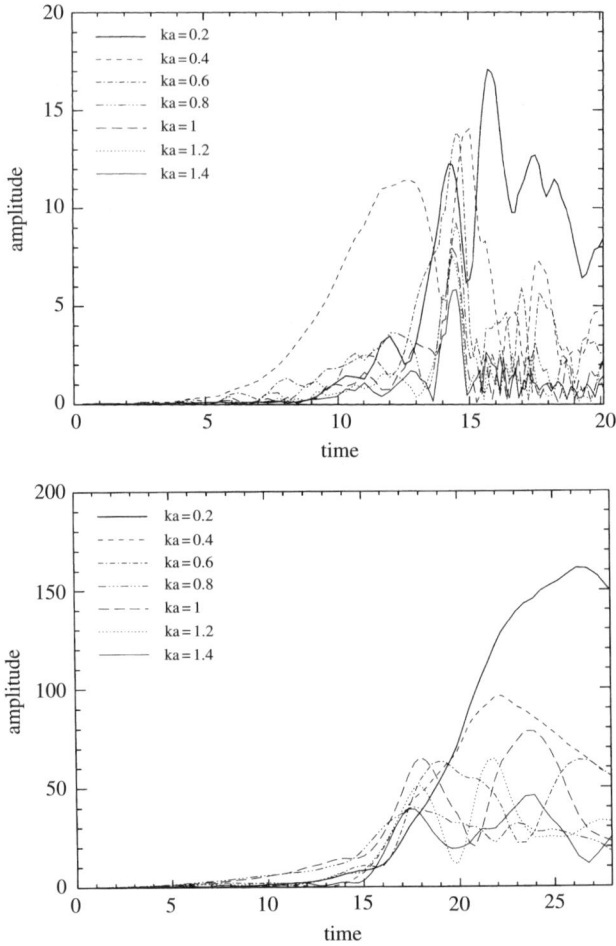

Fig. 10. Plot of the longitudinal amplitude of perturbations vs t for $M = 20$, $\nu = 0.3$ (dense jet, *upper panel*) and $M = 5$, $\nu = 3$ (light jet, *lower panel*): the various lines represent different values of ka

6 Conclusions

It is evident from previous discussions how many astrophysical phenomena may be related to the evolution KHI. For example, the different morphologies observed in extragalactic jets (wiggles, knots, etc.) are very likely related to internal shocks and entrainment processes that are strictly related to the KHI. Then the instability may play a major role in dissipative processes of the bulk kinetic energy of the jet that are different in different classes of extragalactic objects (e.g., high and low power radio galaxies). Also the radiative properties of jets are related to the KHI: in fact relativistic particles may be accelerated

M = 20 $\nu = 0.3$ time = 13.3

M = 20 $\nu = 0.3$ time = 14.9

M = 20 $\nu = 0.3$ time = 17.2

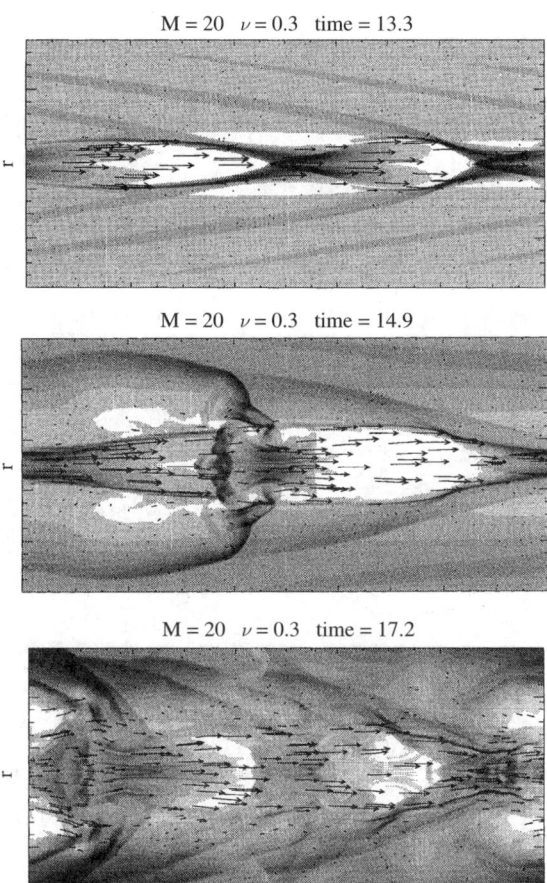

Fig. 11. Maps, for three different values of t, of the density profile (*dark*: high density regions) and of the velocity fields for $M = 20$ and $\nu = 3$ (dense jet)

by a Fermi I process in shocks and by Fermi II interaction with the MHD turbulence fed by the fast evolution of unstable modes with short wavelength. A similar scenario holds also in stellar jets, where the diagnostic of the plasma properties is much easier. Spectroscopic observations of the knots confirm that these are shocks, providing information on the plasma conditions (temperature, density, and velocity), that can be used to constrain the parameters in the numerical simulations in order to consistently reproduce the observed phenomenology.

Efforts to improve the numerical codes and the computing facilities for a more sophisticated analysis of the instability are ongoing. Different physical ingredients (3-D, magnetic fields, and relativistic velocities) as well as 'companion' zeroth order effects (gravity, general relativity, radiative losses, large scale inhomogeneities, etc.) are being considered.

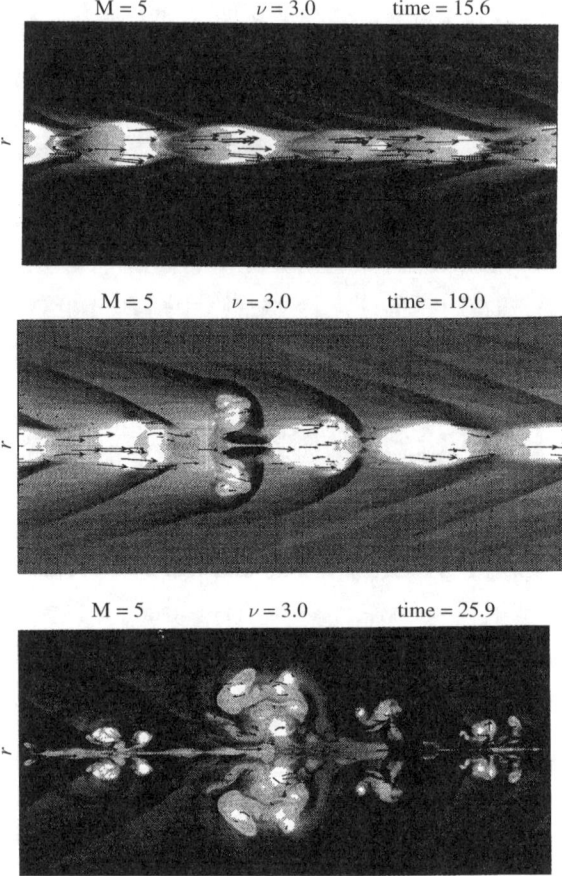

Fig. 12. The same as in Fig. 11 for a light jet ($M = 5$, $\nu = 0.3$)

Since the historical papers by Von Helmholtz and Lord Kelvin [1, 2], numerous papers have been published on the various properties of the KHI: The summary of the fundamental principles may be found in some classical textbooks as Chandrasekhar and Drazin & Reid [3, 4]. The beam model of Rees [5] on radio galaxies extended the interest of the KHI to an astrophysical context with the stability studies of by Blandford and Pringle and Turland and Sheuer [6, 7]. The first numerical simulations on the nonlinear evolution of the KHI were carried on by Hardee and Norman [8, 9]. In the following decades tens of papers have been published on this argument of research, where classical and relativistic flows were considered including magnetic fields and shear layers. When the astrophysical application of the KHI was extended to stellar jets, other effects had to be considered, e.g., radiative losses [10, 11]. A summary and a discussion of the main results (with references therein) can be found in [12, 13, 14].

Concerning in particular the arguments presented here, details on the stability of a planar vortex sheet (Sect. 4.1) are found in [15, 16], while the effects of a shear layer (Sect. 4.2) are discussed in [17, 18, 19, 20]. For the stability of cylindrical jets (Sect. 4.3), we can refer mainly to [21, 22, 23, 24, 25], while the properties of relativistic plasmas (Sect. 4.4) are discussed in the textbooks of Synge and Landau and Lifshitz [26, 27]. Concerning the nonlinear evolution of the KHI presented in Sect. 5, more details are reported in [28, 29].

I wish to conclude by remarking that, in spite of the detailed and sophisticated numerical solutions obtained in these last years regarding jet dynamics and evolution, linear studies of the KHI have been very recently reconsidered, mainly concerning the KHI properties in magnetized, relativistic regime [14, 30, 31, 32, 33].

References

1. von Helmholtz, H., PhMa, **36** (4), 337 (1868)
2. Kelvin, L., PhMa, **42** (4), 362 (1871)
3. Chandrasekhar, S., Hydrodynamic and Hydromagnetic Stability, Oxford University Press, Oxford (1961)
4. Drazin, P.G., & Reid, W.H., Hydrodynamic Stability, Cambridge University Press, Cambridge (1981)
5. Rees, M.J., Nature **229**, 312 (1971)
6. Blandford, R.D., & Pringle, J.E., MNRAS, **176**, 443 (1976)
7. Turland, B.D., & Sheuer, P.A.G., MNRAS, **176**, 421 (1976)
8. Hardee, P.E., & Norman, M.L., AJ, **334**, 70 (1988)
9. Norman, M.L., & Hardee, P.E., AJ, **334**, 80 (1988)
10. Massaglia, S., Trussoni, E., Bodo, G., Rossi, P., & Ferrari, A., AA, **260**, 243 (1992)
11. Bodo, G., Massaglia, S., Rossi, P., Trussoni, E., & Ferrari, A., PFA **5**, 405 (1993)
12. Birkinshaw, M., The stability of jets. In: Beams and Jets in Astrophysics, Hughes, P.A. (Editor). Cambridge University Press, Cambridge, pp. 278–341 (1991)
13. Ferrari, A., ARAA, **36**, 539 (1998)
14. Perucho, M., PhD Thesis, Univ. of Valencia (2005)
15. Gerwin, R.A., RMP, **40**, 652 (1968)
16. Ferrari, A., Trussoni, E., & Zaninetti, L., MNRASoc. **193**, 469 (1980)
17. Drazin, P.G., & Davey, A., JFM, **82**, 255
18. Ray, T.P., MNRAS, **198**, 617 (1982)
19. Ferrari, A., Trussoni, E., MNRAS, **205**, 515 (1983)
20. Birkinshaw, M., MNRAS, **208**, 887 (1984)
21. Ferrari, A., Trussoni, E., & Zaninetti, L., AA, **64**, 43 (1978)
22. Ferrari, A., Trussoni, E., & Zaninetti, L., MNRAS, **196**, 1054 (1981)
23. Hardee, P.E., AJ, **234**, 47 (1979)
24. Hardee, P.E., AJ, **257**, 509 (1982)
25. Ray, T.P., MNRAS, **196**, 195 (1981)

26. Synge, J.L., The Relativistic Gas, North Holland Publishing Company, Amsterdam (1957)
27. Landau, L.D., & Lifshitz, E.M., Fluid Mechanics, Chap. 15, Pergamon Press, Oxford (1959)
28. Trussoni, E., Bodo, G., Ferrari, A., & Massaglia, S., MSAI **64**, 87 (1993)
29. Bodo, G., Massaglia, S., Ferrari, A., & Trussoni, E., AA, **283**, 655 (1994)
30. Bodo, G., Mignone, A., & Rosner, R. PRE **70**, 6304 (2004)
31. Mignone, A., McKinney, J.C., MNRAS, **378**, 1118–1130 (2007)
32. Osmanov, Z., PhD Thesis, Univ. of Torino (2007)
33. Perucho, M., Hanasz, M., Marti, J.M., & Miralles, J.A., PRE **75**, 6312–6321 (2007)

Pressure-Driven Instabilities
in Astrophysical Jets

P.-Y. Longaretti

Laboratoire d'Astrophysique de Grenoble (LAOG), CNRS and Université
Joseph Fourier, BP 53 38041 Grenoble Cedex 9, France,
Pierre-Yves.Longaretti@obs.ujf-grenoble.fr

Abstract Astrophysical jets are widely believed to be self-collimated by the hoop-stress due to the azimuthal component of their magnetic field. However this implies that the magnetic field is largely dominated by its azimuthal component in the outer jet region. In the fusion context, it is well-known that such configurations are highly unstable in static columns, leading to plasma disruption. It has long been pointed out that a similar outcome may follow for MHD jets, and the reasons preventing disruption are still not elucidated, although some progress has been accomplished in the recent years.

Longaretti, P.-Y.: *Pressure-Driven Instabilities in Astrophysical Jets.* Lect. Notes Phys. **754**,
131–151 (2008)
DOI 10.1007/978-3-540-76967-5_4   © Springer-Verlag Berlin Heidelberg 2008

In these notes, I review the present status of this open problem for pressure-driven instabilities, one of the two major sources of ideal MHD instability in static columns (the other one being current-driven instabilities).

I first discuss in a heuristic way the origin of these instabilities. Magnetic resonances and magnetic shear are introduced, and their role in pressure-driven instabilities discussed in relation to Suydam's criterion. A dispersion relation is derived for pressure-driven modes in the limit of large azimuthal magnetic fields, which gives back the two criteria derived by Kadomtsev for this instability. The growth rates of these instabilities are expected to be short in comparison with the jet propagation time.

What is known about the potential stabilizing role of the axial velocity of jets is then reviewed. In particular, a nonlinear stabilization mechanism recently identified in the fusion literature is discussed.

Keywords Ideal MHD: stability · pressure-driven modes · Jets: stability

1 Introduction

The enormous distances over which astrophysical jets propagate without losing their coherence certainly constitute one of the most striking features of these objects. Typically, jets from Young Stellar Objects (YSOs) do reach out to a few parsecs, while the radial extent of their region of origin appears to be smaller than ~ 100 A.U, making jets extremely elongated structures.

Blandford and Rees [3] already pointed out, in their 1974 pioneering work, that in laboratory experiments, jets do not propagate much farther than about ten times their radii, which makes the propagation lengths of astrophysical jets all the more impressive. The simplest way out of this conundrum would be to assume that jets are ballistic. Indeed, for YSO jets at least, the observed opening angle ($\sim 5°$) is consistent with the idea that they freely expand when one compares their thermal and bulk velocities. However, this option leaves open the issue of the formation of such powerful jets in the first place. In addition, as critically, the ballistic hypothesis does not explain how these jets survive the development of the Kelvin–Helmholtz instability, which is now known to be quite disruptive in purely hydrodynamic jets [4] [5].

The shortcomings of the simple ballistic picture certainly motivated the elaboration of MHD jet models to some extent. Such models, however, are also prone to instabilities. The most important ones discussed in the literature can be grouped into three categories:

- MHD Kelvin–Helmholtz instability. As for its HD counterpart, the driving agent is the velocity gradient at the jet/external medium interface. This instability has received a lot of attention in the literature, as the largest source of free energy in a jet is its bulk motion.
- Conversely, the presence of a magnetic field provides a source of instability even in the absence of bulk motion. Ideal MHD instabilities are commonly

divided into current- and pressure-driven according to the driving factor (equilibrium parallel current in the first case and gas pressure versus field-line curvature in the second).

- Radiative instabilities related to the coupling of the radiation field and of the plasma dynamical quantities.

The structure of jets is not precisely known, which is one of the difficulties in analyzing their stability. Most stability analyses assume that jets can be described as some sort of cylindrical column in motion, pervaded by a magnetic field. Self-collimated jet models are not exactly cylindrical, but as the observed opening angles are small, the assumption of cylindrical shape is not expected to be a major limitation.

More critically, such jet models have a helical field structure with the azimuthal component of the magnetic field dominating over the vertical one in the outer jet regions. This follows in most models because the magnetic tension is the confining force ensuring self-collimation. Static MHD columns (i.e., not subject to the bulk motion characterizing MHD jets) pervaded by a helical magnetic field are referred to as "screw pinches" in the fusion literature. It is also known in this context that the dominance of the azimuthal field component leads to both types of MHD instabilities mentioned above and may cause the disruption of the plasma column itself on a few dynamical time-scales. This has long been an argument against magnetically self-confined jet models. However, recent investigations indicate that a bulk motion can play an important stabilizing role (see Sect. 5). Conversely, the presence of a magnetic field can help stabilizing the Kelvin–Helmholtz modes [20]. These recent advances seem to indicate that a sophisticated equilibrium jet structure is required if one is to understand jet stability properties, a state of development not yet reached by the subject, but that now appears to be within sight.

To conclude these introductory remarks, I would like to point out that, in the nonlinear phase, an instability can have three broad types of outcome: (i) disruption of the fluid configuration (in the case, at hand, of the jet as a jet); (ii) internal reorganization, the flow becoming laminar again in the end; and (iii) turbulence (with or without internal reorganization of the structure). The most prominent objective of the study of jet stability is to understand how the first issue is avoided in real jets; this issue may well be seen as our inability to formulate the initial value problem correctly. A second but important issue is to understand how turbulence might be driven by jet destabilization. This issue is probably more important in AGN than in YSO jets, as turbulence is often invoked in the former context as a source of high energy particle acceleration.

The object of these lecture notes is pressure-driven instabilities. As most investigations of this problem have been made in the fusion context for static columns, this essential aspect of the subject will first be reviewed before briefly presenting the more recent (and more scant) results on moving columns. The next section presents some general ideas about the physical origin of MHD instabilities; the concept of magnetic shear is introduced there as well, and its stabilizing role on pressure-driven instabilities, expressed by Suydam criterion, is discussed. Section 3 introduces the Lagrangian form of the perturbation equations used in static columns, as this formulation is the most powerful to derive general results, such as those derived from the Energy Principle presented in Sect. 4. Section 5 presents the dispersion relation of pressure-driven instabilities in low magnetic shear that are expected to characterize jet's outer regions. The few published results on moving columns (i.e., jets) are presented in Sect. 6. The last section summarizes the present state of understanding of this aspect of jet stability and outlines areas where improvement is needed.

Sections 2 and 6 are intended for a general audience, while Sects. 3, 4 and 5 are more theoretical in nature. The exposition is aimed at the graduate student level.

2 Heuristic Description of MHD Instabilities

This first section is intended to provide the reader with some qualitative and semi-quantitative ideas about the onset and characteristics of pressure-driven instabilities, leaving technical aspects of the stability analysis to the later sections.

2.1 Qualitative Conditions of Ideal MHD Instabilities in Static Equilibria

The equilibrium configurations leading to an ideal MHD instability have been well investigated in the fusion literature. For current-driven instabilities, the first criterion was devised by Kruskal and Shafranov. Basically, it states that, in cylindrical column of length L, instability follows if the magnetic field line rotates more than a certain number of times around the cylinder from end to end. The exact number of rotations required for instability is dependent on the considered equilibrium configuration; it is usually of order unity.

Concerning pressure-driven instabilities, a more clear-cut necessary condition of instability can be stated: instability follows once the pressure force pushes the plasma outwards from the inside of the field line curvature. This condition can be derived from the Energy Principle, as will be shown later on.

These conditions of onset of instability are illustrated on Fig. 1. In an actual plasma, the origin of an instability (current- or pressure-driven) is usually not easy to pinpoint except in special instances. For example, if the plasma is cold (no pressure force), the instability is necessarily current-driven. Also the growth rates of current-driven modes are known to decrease with spatial order – e.g., they decrease with increasing azimuthal wavenumber m – while the most unstable pressure-driven modes have a growth rate which is nearly independent on the wavenumber. Consequently, large wave, number unstable modes are therefore always pressure-driven in a static, ideal MHD

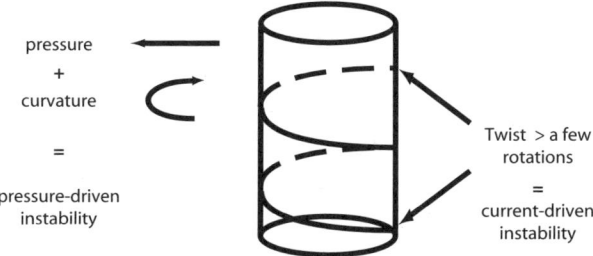

pressure
+
curvature

=

pressure-driven
instability

Twist > a few
rotations

=
current-driven
instability

Fig. 1. Qualitative description of the conditions of onset of MHD instabilities (see text for details)

column. Besides these two limiting cases, an MHD instability almost always results from an inseparable mix of pressure and current driving. If the column is moving, the distinction between Kelvin–Helmholtz, current- and pressure modes is even more blurred, except in some cases, where branches of instability can be identified by taking appropriate limits.

In terms of outcome of the instability, it is essential to know whether unstable modes are internal or external, i.e., they have vanishing or substantial displacement on the plasma surface (here, the jet surface). It is well known in the fusion context that unstable external modes are prone to disrupt the plasma, as may be the case, e.g., with the $m = 1$ ("kink") current-driven mode.

In the next sections, mostly high wavenumber modes will be examined, where the pressure driving is most obvious, in order to best identify the characteristic features of this type of instability.

2.2 Magnetic Shear, Magnetic Resonances, and Suydam's Criterion

The concept of magnetic shear plays an important role in the understanding of the stability of pressure-driven mode. The magnetic shear characterizes the change of orientation of field lines when moving perpendicularly to magnetic surfaces. In the case of cylindrical equilibria, this concept is illustrated on Fig. 2. Magnetic surfaces are cylindrical. Field lines within magnetic surfaces have a helix shape; the change of helix pitch rB_z/B_θ characterizes the magnetic shear. A quantity related to the pitch and largely used in the fusion community is the safety factor q:

$$q = \frac{rB_z}{R_o B_\theta},\tag{1}$$

where R_o is the column radius. For reasons to be discussed later, a high enough safety factor is required for stability, hence its name. The magnetic shear s is defined as

$$s \equiv \frac{r}{q}\frac{\mathrm{d}q}{\mathrm{d}r}.\tag{2}$$

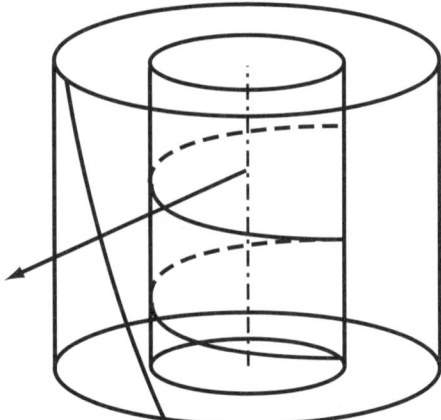

Fig. 2. The change in the pitch of field lines between magnetic surfaces is the source of the magnetic shear (see text)

Magnetic resonances constitute another important key to the question of stability. A cylindrically symmetric equilibrium is invariant in the vertical and azimuthal direction, so that perturbations from equilibrium can, without loss of generality, be expanded in Fourier terms in these directions and assumed to be proportional to $\exp i(m\theta + kz)$. Magnetic resonances are the (cylindrical) surfaces where the wavevector $\boldsymbol{k} = m/r\boldsymbol{e}_\theta + k\boldsymbol{e}_z$ is perpendicular to the equilibrium magnetic field:

$$\frac{\boldsymbol{k} \cdot \boldsymbol{B}_o}{B_o} \equiv k_\parallel = \frac{1}{B_o}\left(\frac{m}{r}B_\theta + kB_z\right) = 0 \qquad (3)$$

where $k_\parallel$ is the component of the wavevector parallel to the equilibrium magnetic field. The significance of these surfaces stems from the fact that, in general, dispersion relations incorporate a stabilizing piece of the form $V_A^2 k_\parallel^2$, where V_A is the Alfvén speed. This term is responsible for the propagation of Alfvén waves and arises from the restoring force due to the magnetic tension (see Sect. 5 for the precise meaning of these statements). As such, it is always stabilizing. Obviously, this stabilization is minimal in the vicinity of a magnetic resonance for a given (m, k) mode, so that pressure-driven instabilities are preferentially triggered at magnetic resonances for any given mode.

Note, however, that a large magnetic shear limits the role of magnetic resonances in the destabilization of the plasma. Indeed, defining the perpendicular wavenumber

$$k_\perp = -\frac{1}{B_o}\left(\frac{m}{r}B_z - kB_\theta\right), \qquad (4)$$

and designating by r_c, the radial position of the magnetic resonance of the (m, k) mode, one finds that

$$k_\parallel \simeq \frac{B_\theta B_z}{B_o^2}k_\perp s\frac{r - r_c}{r_c}, \qquad (5)$$

to first order in $(r - r_c)/r_c$ in the vicinity of the magnetic resonance r_c. This implies that $V_A k_\parallel$ will remain small either if $s \ll 1$ (small shear) or if the magnetic field is mostly perpendicular ($|B_\theta| \ll |B_z|$) or azimuthal ($|B_z| \ll |B_\theta|$) so that $|B_\theta B_z|/B_o^2 \ll 1$. However, if the field is predominantly vertical, it is little curved, and pressure destabilization is expected to be weak or nonexistent according to the description of the condition on instability depicted in Fig. 1; furthermore, s being a logarithmic derivative is usually of order unity. Therefore, in practice, stabilization by magnetic tension will be reduced essentially when the field is mostly azimuthal.

These features are embodied in Suydam criterion, which expresses a sufficient condition for instability:

$$\frac{B_z^2}{8\mu_o r} s^2 + \frac{dP}{dr} > 0. \tag{6}$$

The converse of this statement is a necessary condition for stability. The origin of this criterion is briefly discussed in Sect. 4. It turns out that this condition is both a necessary and sufficient condition of instability for large wavenumber modes [10]. The condition of instability requires $dP/dr < 0$, which agrees with our heuristic description of the onset of instability given above. It will also be discussed in Sect. 4 that the growth rates γ of pressure-driven instabilities are $\gamma \sim C_S/R_o$ (C_S is the sound speed and R_o the jet radius).

Coming back to Eq. (6), the first term is stabilizing, but the stabilization will be minimal in the condition just discussed, i.e., when the field is mostly azimuthal. Indeed, in this case, the equilibrium condition Eq. (11) implies that $dP/dr \sim B_\theta^2/\mu_o r \gg B_z^2/r \sim B_z^2 s^2/r$. This situation is expected to hold in magnetically self-confined jet's outer regions. Indeed, most such jet models (e.g., [19] and [11]) have $|B_\theta| \gg |B_z|$ in the asymptotic jet regime to ensure confinement. This feature combined to the previous statement that MHD instabilities involving the boundary are most prone to disrupt static MHD columns makes the assessment of the role of pressure-driven instabilities in MHD jets particularly critical for the viability of such models. This viability hinges on the hopefully stabilizing role of the jet bulk motion (see Sect. 6).

3 Ideal MHD in Static Columns

The simplest framework, in which the stability of jets can be investigated, is ideal magnetohydrodynamics (MHD). Justifications and limitations of this approach are briefly discussed in Appendix A.

3.1 Equations

The MHD equations used in these notes are the continuity equation, the momentum equation without the viscous term, the induction equation without the resistive term, and a polytropic equation of state. Incompressibility is not assumed, as pressure-driven modes are not incompressible except at the marginal stability limit. These equations read

$$\frac{\partial \rho}{\partial t} + \nabla \rho \boldsymbol{v} = 0, \tag{7}$$

$$\frac{\partial \boldsymbol{v}}{\partial t} + \boldsymbol{v} \cdot \nabla \cdot v = -\frac{\nabla P_T}{\rho} + \frac{\boldsymbol{B} \cdot \nabla \boldsymbol{B}}{\mu_o \rho}, \tag{8}$$

$$\frac{\partial \boldsymbol{B}}{\partial t} = \nabla \times (\boldsymbol{v} \times \boldsymbol{B}), \tag{9}$$

$$P = K \rho^\gamma, \tag{10}$$

with standard notations, where $P_T = P + B^2/2\mu_o$ is the total (gas and magnetic) pressure, K a constant, and γ is the polytropic index.

3.2 Equilibrium

Using a cylindrical coordinate system (r, θ, z), a static ($\mathbf{v} = 0$) cylindrical column of axis z is described by a helical magnetic field $B_\theta(r), B_z(r)$, and a gas pressure $P(r)$ depending only on the cylindrical radius r. The continuity and induction equations as well as the vertical and azimuthal component of the momentum equation are then trivially satisfied, while the radial component reduces to

$$-\frac{\mathrm{d}P_T}{\mathrm{d}r} - \frac{B_\theta^2}{\mu_o r} = 0. \tag{11}$$

This cylindrical equilibrium is best characterized by introducing a number of quantities homogeneous to an inverse length, both in vectorial ($\boldsymbol{\mathcal{K}_B}, \boldsymbol{\mathcal{K}_P}$ and $\boldsymbol{\mathcal{K}_C}$) or algebraic form ($\mathcal{K}_B, \mathcal{K}_P$ and $\mathcal{K}_C$). They are defined by:

$$\boldsymbol{\mathcal{K}_B} \equiv \frac{\boldsymbol{\nabla} B_o}{B_o} = \frac{1}{B_o} \frac{\mathrm{d}B_o}{\mathrm{d}r} \mathbf{e}_r \equiv \mathcal{K}_B \mathbf{e}_r, \tag{12}$$

$$\boldsymbol{\mathcal{K}_P} \equiv \frac{\boldsymbol{\nabla} P_o}{P_o} = \frac{1}{P_o} \frac{\mathrm{d}P_o}{\mathrm{d}r} \mathbf{e}_r \equiv \mathcal{K}_P \mathbf{e}_r, \tag{13}$$

$$\boldsymbol{\mathcal{K}_C} \equiv \mathbf{e}_\parallel \cdot \boldsymbol{\nabla} \mathbf{e}_\parallel = -\frac{B_\theta^2}{r B_o^2} \mathbf{e}_r \equiv \mathcal{K}_C \mathbf{e}_r. \tag{14}$$

where B_o and P_o are the equilibrium distribution of magnetic field and gas pressure, and $\mathbf{e}_\parallel = \boldsymbol{B}_o/B_o$ is the unit vector parallel to the magnetic field; $\boldsymbol{\mathcal{K}_C}$ is the curvature vector of the magnetic field lines, and $\boldsymbol{\mathcal{K}_B}$ characterizes the inverse of the spatial scale of variation of the magnetic field, while $\boldsymbol{\mathcal{K}_P}$ characterizes the inverse scale of variation of the fluid pressure. The first identity in these relations is general, whereas the second one pertains to cylindrical equilibria only.

It is also convenient to introduce the plasma β parameter:

$$\beta \equiv \frac{2\mu_o P_o}{B_o^2}, \tag{15}$$

This parameter measures the relative importance of the gas and magnetic pressures.

With these definitions, the jet force equilibrium relation reads

$$\frac{\beta}{2}\mathcal{K}_P = (\mathcal{K}_C - \mathcal{K}_B), \tag{16}$$

Both forms of the equilibrium relation, Eqs. (11) and (16), express the fact that the hoop stress due to the magnetic tension $(\mathcal{K}_C)$ balances the gas $(\mathcal{K}_P)$ and magnetic $(\mathcal{K}_B)$ pressure gradient to achieve equilibrium and confine the plasma in the column. Self-confinement is achieved in this way when the external pressure is negligible at the column boundary.

3.3 Perturbations

We want to investigate the stability with respect to deviations from equilibrium. As the background equilibrium is static, the problem is most easily formulated and analyzed in Lagrangian form: indeed, in this case, all equations but the momentum equation can be integrated with respect to time. To this effect, we introduce, for any fluid particle at position $\boldsymbol{r}_o$ in the absence of perturbation, the displacement $\boldsymbol{\xi}(\boldsymbol{r}_o, t)$ at time t from its unperturbed position, so that its actual position is given by

$$\boldsymbol{r}(\boldsymbol{r}_o, t) = \boldsymbol{r}_o + \boldsymbol{\xi}(\boldsymbol{r}_o, t). \tag{17}$$

The unperturbed position $\boldsymbol{r}_o$ is used to uniquely label all fluid elements.

Denoting by δX, the (Lagrangian) variation during the displacement of any quantity X, the linearized (Eulerian) equation of continuity $\partial_t \delta\rho = -\nabla(\rho_o \boldsymbol{v})$ integrates into[1]

$$\delta\rho = -\nabla(\rho_o \boldsymbol{\xi}). \tag{18}$$

Similarly, the linearized induction equation $\partial_t \delta\boldsymbol{B} = \nabla \times (\boldsymbol{B}_o \times \boldsymbol{v})$ leads to

$$\delta\boldsymbol{B} = \nabla \times (\boldsymbol{B}_o \times \boldsymbol{\xi}). \tag{19}$$

From these results and the polytropic equation of state, the total pressure variation reads

---

[1] In these expressions, the difference between the Eulerian and Lagrangian variations has been ignored as they disappear to first order in the displacement $\boldsymbol{\xi}$ in the final equations. For the same reason, no distinction is made between the derivative with respect to $\boldsymbol{r}$ or $\boldsymbol{r}_o$.

$$\delta P_T = -\boldsymbol{\xi} \cdot \nabla P_o - \gamma P_o \nabla \cdot \boldsymbol{\xi} + \frac{\boldsymbol{B}_o \cdot \delta \boldsymbol{B}}{\mu_o}. \tag{20}$$

For static equilibria, one can, without loss of generality, take a Fourier transform of the linearized momentum equation with respect to time. For a given Fourier mode, one can write $\boldsymbol{\xi}(\boldsymbol{r}, t) = \boldsymbol{\xi}(r) \exp i\omega t$, so that the linearized momentum equation becomes

$$-\rho_o \omega^2 \boldsymbol{\xi} = -\nabla \delta P_T + \delta \boldsymbol{T} \equiv \boldsymbol{F}(\boldsymbol{\xi}), \tag{21}$$

where $\delta \boldsymbol{T} = (\boldsymbol{B}_o . \nabla \boldsymbol{B} + \boldsymbol{B} . \nabla \boldsymbol{B}_o)/\mu_o$ represents the variation of the magnetic tension force.[2] The last identity in Eq. (21) defines the linear operator $\boldsymbol{F}$, operating on $\boldsymbol{\xi}$ through Eqs. (19) and (20).

4 The Energy Principle and Its Consequences

The linear operator $\boldsymbol{F}$ of Eq. (21) is self-adjoint, i.e., taking into account that $\boldsymbol{F}$ is real:

$$\int \boldsymbol{\eta} \cdot \mathbf{F}(\boldsymbol{\xi}) d^3 r = \int \boldsymbol{\xi} \cdot \mathbf{F}(\boldsymbol{\eta}) d^3 r. \tag{22}$$

A demonstration of this relation can be found, e.g., in Freidberg [12] (cf p. 242 and Appendix A of the book).

As a consequence of this property of $\boldsymbol{F}$, an Energy Principle can be formulated. Defining

$$\delta W(\boldsymbol{\xi}^*, \boldsymbol{\xi}) = -\frac{1}{2} \int \boldsymbol{\xi}^* \cdot \mathbf{F}(\boldsymbol{\xi}) d^3 r, \tag{23}$$

and

$$K(\boldsymbol{\xi}^*, \boldsymbol{\xi}) = \frac{1}{2} \int \rho |\boldsymbol{\xi}|^2 d^3 r, \tag{24}$$

and taking the scalar product of Eq. (21) with $\boldsymbol{\xi}^*$ leads to

$$\omega^2 = \frac{\delta W}{K}. \tag{25}$$

The self-adjointness of $\mathbf{F}$ has two important consequences (Energy Principle):

1. ω^2 is also extremum with respect to a variation of $\boldsymbol{\xi}$.
2. Stability follows if and only if $\delta W > 0$ for all possible $\boldsymbol{\xi}$.

Ascertaining stability through the last statement is usually an impossible task. Instead, one usually makes use of the Energy Principle in a less ambitious manner: if one can find some displacement making $\delta W < 0$ then one has a sufficient condition of instability (or, taking the converse statement, a necessary condition of stability). This is actually how Suydam criterion is

---

[2] Within a factor ρ_o.

demonstrated. First the expression of δW is simplified by taking advantage of the cylindrical geometry and focusing on marginal stability and incompressible displacements (as they make δW more easily negative; see below). Next, one chooses a particular form of displacement in the vicinity of the magnetic resonance of an (m, k) mode, and looks under which conditions this displacement makes δW negative; the condition turns out to be Suydam criterion for a well-chosen displacement. These computations are rather lengthy and the reader is referred to Freidberg's book [12] for details.

A useful form[3] of δW has been derived by Bernstein et al. [2], which reads (see Freidberg [12], p. 259)

$$
\delta W = \frac{1}{2} \int d^3 r \left[\frac{|\boldsymbol{Q}_\perp|^2}{\mu_o} + \frac{B_o^2}{\mu_o} |\nabla \cdot \boldsymbol{\xi}_\perp + 2\boldsymbol{\mathcal{K}_C} \cdot \boldsymbol{\xi}_\perp|^2 + \gamma P_o |\nabla \cdot \boldsymbol{\xi}|^2 \right.
$$
$$
\left. - 2P_o (\boldsymbol{\mathcal{K}_P} \cdot \boldsymbol{\xi}_\perp)(\boldsymbol{\mathcal{K}_C} \cdot \boldsymbol{\xi}_\perp^*) - J_\| (\boldsymbol{\xi}_\perp^* \times \boldsymbol{e}_\|) \cdot \boldsymbol{Q}_\perp \right], \tag{26}
$$

where $\boldsymbol{\xi}_\perp$ is the component of the displacement perpendicular to the unperturbed field $\boldsymbol{B}$, $\boldsymbol{Q}_\perp = \nabla \times (\boldsymbol{\xi}_\perp \times \boldsymbol{B}_o)$ is the perturbation in the magnetic field, $\boldsymbol{\mathcal{K}_C}$ is the curvature vector of the magnetic field, and $\boldsymbol{\mathcal{K}_P}$ is the inverse pressure length-scale vector defined earlier; $J_\|$ and $\boldsymbol{e}_\|$ are the current and unit vector parallel to the magnetic field, respectively. The quantities $\boldsymbol{\mathcal{K}_P}$ and $\boldsymbol{\mathcal{K}_C}$ are defined in Eqs. (13) and (14).

The first term describes the field line bending energy; it is the term responsible for the propagation of Alfvén waves, through the restoring effect of the magnetic tension, which makes field lines acting somewhat like a rubber band. The second term is the energy in the field compression, while the third is the energy in the plasma compression. The fourth term arises from the perpendicular current (as $\nabla P = \boldsymbol{J}_\perp \times \boldsymbol{B}$ in a static equilibrium), and the last one arises from the parallel current $J_\|$. Only these two terms can be negative and give rise to an instability if they are large enough to make $\omega^2 < 0$. Pressure-driven instabilities are driven by the first of these two terms, while current-driven instabilities are due to the second one. Pressure-driven instabilities are further subdivided into interchange and ballooning modes, depending on the shape of the perturbation, but the basic properties of these different modes are similar, and this distinction will not be discussed further in these notes.[4]

For our purposes here, we are mostly interested in what can be learned from the form of the fourth term. First note that this term is destabilizing in cylindrical geometry when $\mathcal{K}_C \mathcal{K}_P > 0$; this justifies the necessary condition of instability given in Sect 2.1. Furthermore, Eqs. (25) and (26) imply that the pressure-driving term produces an inverse growth rate γ, of the order of magnitude

$$
|\gamma|^2 \sim C_S^2 \mathcal{K}_C \mathcal{K}_P \sim C_S^2 / R_o^2, \tag{27}
$$

---

[3] The boundary term is ignored, as it is not required in this discussion.

[4] In particular, Suydam criterion applies also to ballooning mode in cylindrical geometry; see Freidberg's book [12], pp. 401–402 for details.

where C_S is the sound speed and R_o the column radius. This is quite fast, comparable to the Kelvin–Helmholtz growth rate in YSO jets. This order of magnitude will be used in the next section to set up an ordering leading to analytically tractable dispersion relations for pressure-driven unstable modes.

5 Dispersion Relation in the Large Azimuthal Field Limit

Most of the general results on pressure-driven instabilities were obtained in the fusion literature either from the use of the Energy Principle or from the so-called Hain-Lüst equation (a reduced perturbation equation for the radial displacement [15] [13]). These approaches are quite powerful but not familiar to the astrophysics community, and involve a lot of prerequisite.

It is more common in astrophysics to grasp the properties of an instability through the derivation of a dispersion relation. There are actually two papers doing this in the jet context for pressure-driven instabilities; however, the first one, by Begelman [1], focuses on the relativistic regime which brings a lot of added complexity to the discussion, and the second one [17] is partially erroneous.

Fortunately, in the limit of a near toroidal field of interest here, a dispersion relation can be derived ab initio by elementary means, and this approach is adopted here. To this effect, it is first useful to reexamine the behavior of the three MHD modes in an homogeneous medium, in the limit of quasiperpendicular propagation. It is known that this limit allows the use of a kind of WKB type of approach in the study of interchange and ballooning pressure-driven modes (see, e.g., Dewar and Glasser [9]), a feature we shall take advantage of in these notes.

5.1 MHD Waves in Quasi-perpendicular Propagation in Homogeneous Media

We consider an homogeneous medium pervaded by a constant magnetic field $\boldsymbol{B}_o$. The analysis of linear perturbations in such a setting leads to the well-known dispersion relation of the slow and fast magnetosonic modes and the Alfvén mode. Our purpose here is to point out useful features of these modes when the wavevector is nearly perpendicular to the unperturbed magnetic field.

To this effect, let us consider plane wave solutions to Eq. (21), where $\xi \propto \exp(-i\boldsymbol{k} \cdot \boldsymbol{r})$, and assume that the direction of propagation is nearly perpendicular to the magnetic field, i.e., $k_\parallel \ll k_\perp$ (defined in Eqs. (5) and (4)). The focus on quasi perpendicular propagation comes from the remarks of Sect. 2.2, where it was noted that instability is easier to achieve in the vicinity of magnetic resonances, i.e., where $k_\parallel \ll k_\perp$.

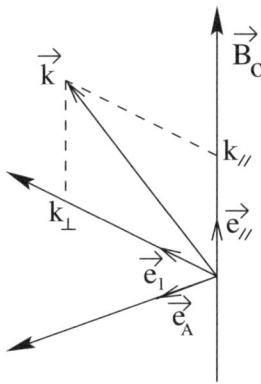

Fig. 3. Definition of the reference frame $(e_\parallel, e_l, e_A)$

Let us also introduce the orthogonal reference frame $(e_\parallel, e_l, e_A)$ where $e_\parallel \equiv B_o/B_o$ is parallel to the unperturbed magnetic field, $e_l \equiv k_\perp/k_\perp$, and $e_A \equiv e_\parallel \times e_l$ (see Fig. 3). With our definition of $k_\parallel$ and $k_\perp$ in Eqs. (5) and (4), $e_A = e_r$. The subscripts l and A stand for longitudinal and Alfvénic, respectively ($e_\parallel$, e_l, and e_A are the directions of the displacement of purely slow, fast and Alfvénic modes in the limit of nearly transverse propagation adopted here, as shown below).

Denoting $(\xi_\parallel, \xi_l, \xi_A)$ the components of the lagrangian displacement $\boldsymbol{\xi}$ in this reference frame, the momentum equation Eq. (21) yields the following three component equations

$$\left(\omega^2 - C_S^2\, k_\parallel^2\right) \xi_\parallel = C_S^2\, k_\parallel\, k_\perp\, \xi_l, \tag{28}$$

$$\left(\omega^2 - C_S^2\, k_\perp^2 - V_A^2\, k^2\right) \xi_l = C_S^2\, k_\parallel\, k_\perp\, \xi_\parallel, \tag{29}$$

$$\left(\omega^2 - V_A^2\, k_\parallel^2\right) \xi_A = 0, \tag{30}$$

while the total pressure perturbation becomes

$$\delta P_T = -i\rho_o \left[\left(C_S^2 + V_A^2\right) k_\perp \xi_l + C_S^2\, k_\parallel \xi_\parallel\right]. \tag{31}$$

Equation (30) gives the dispersion relation of Alfvén waves, $\omega_A^2 = V_A^2 k_\parallel^2$, which decouple from the two magnetosonic modes described by the remaining two equations. The solutions of the magnetosonic modes are easily derived and possess the following important properties. Characterizing quasiperpendicular propagation with the small parameter $\epsilon \equiv |k_\parallel/k_\perp| \ll 1$, these two equations imply $\omega_S^2 \simeq C_S^2 V_A^2/(C_S^2 + V_A^2)k_\parallel^2$ and $\xi_l \sim O(\epsilon\xi_\parallel)$ for the slow magnetosonic wave, while $\omega_F^2 \simeq (C_S^2 + V_A^2)k_\perp^2$ and $\xi_\parallel \sim O(\epsilon\xi_l)$ for the fast magnetosonic one.

Furthermore, the ξ_l momentum component Eq. (29) combined with Eq. (30) and the ordering of the displacement component just pointed out implies that $\delta P_T = 0$ to leading order in ϵ for the slow magnetosonic mode; note that the same property holds by construction for the Alfvén mode. The cancellation of the total pressure for these two modes is essential from a technical point of view, and will lead to substantial simplification in the derivation of a dispersion relation performed in the next subsection.

5.2 Dispersion Relation and Kadomtsev Criteria

Let us now come back to cylindrical inhomogeneous equilibria. Remember from Sect. 4 that the pressure-driving term will contribute a destabilizing term $\omega^2 \sim C_S^2/R_o^2$ to the dispersion relation. This term will be able overcome the stabilizing effect of the restoring forces of the Alfvén and slow magnetosonic modes only if $V_A|k_\parallel|$, $C_S|k_\parallel| \lesssim C_S/R_o$. This constraint can be achieved in the vicinity of magnetic resonance as previously noted.

More precisely, a simplified dispersion relation can be found in the WKB limit with a displacement of the form $\boldsymbol{\xi}(\boldsymbol{r}) = \boldsymbol{\xi} \times \exp -i(k_r r + m\theta + k_z z)$, if the following ordering is satisfied:

- $|k_\parallel r| \ll$ or $\lesssim 1 \ll |k_r r| \ll |k_\perp r|$: The first inequality ensures that the stabilization by magnetic tension is ineffective (closeness to a resonance). The following inequalities ensure that a WKB limit can be taken. The implied ordering[5] $|k_\parallel| \ll k_\perp$ ensures that δP^* will vanish to leading order as in the homogeneous case discussed in the previous section. The last inequality allows us to neglect the contribution of the radial gradient of total pressure (which *does not* vanish), and greatly simplifies the analysis.
- $|B_z/B_\theta|^2 s^2|k_\perp| \ll |k_\parallel|$: This limit, which applies when $|B_\theta| \gg |B_z|$, ensures that the magnetic shear is not stabilizing.
- $|\omega^2| \ll |\omega_F|^2$: This excludes the fast mode from the problem in the near perpendicular propagation regime considered here. As the fast mode is not expected to be destabilized in this regime (as $|\omega_F^2| \gg V_A^2/r^2$), this does not limit the generality of the results while simplifying the analysis.

It turns out that the resulting dispersion relation captures most of the physics of pressure-driven instabilities; this follows because the most unstable modes have growth rates nearly independent of the azimuthal wavenumber m [10], and because the current-driven instabilities are efficient only at low m and disappear from a WKB analysis.

As previously, the projection Eq. (21) on the longitudinal direction $\boldsymbol{e}_l$ shows that the total pressure perturbation vanishes and that $\xi_l \sim |k_\parallel/k_\perp|\xi_\parallel \ll \xi_\parallel)$, while the components in the other two directions $(\boldsymbol{e}_\parallel, \boldsymbol{e}_r)$ are now coupled

---

[5] For consistency with the previous sections, $k_\perp$ is the wavenumber in the longitudinal direction; it does not include the piece in the radial direction.

and read (some details of the derivation of these equations can be found in Appendix B)

$$\left(\omega^2 - V_{SM}^2 k_{\parallel}^2\right) = -i\frac{2\beta^*}{1+\beta^*}V_A^2\mathcal{K}_C k_{\parallel}\xi_r, \tag{32}$$

$$\left[\omega^2 - V_A^2(k_{\parallel}^2 + k_o^2)\right] = i\frac{2\beta^*}{1+\beta^*}V_A^2\mathcal{K}_C k_{\parallel}\xi_{\parallel}, \tag{33}$$

where $\beta^* = C_S^2/V_A^2$, and $V_{SM}^2 = C_S^2 V_A^2/(C_S^2 + V_A^2)$ is the slow mode speed in the near perpendicular propagation limit. The coupling of the modes blurs their character except in limiting cases.

The quantity k_o^2 is defined as

$$k_o^2 = \frac{4\beta^*}{1+\beta^*}\mathcal{K}_C^2 - 2\beta^*\mathcal{K}_C\mathcal{K}_\rho. \tag{34}$$

Note that if $\mathcal{K}_C = 0$ (i.e., when reverting to an homogeneous medium), Eqs. (32) and (33) yield back the slow and Alfvén mode, respectively. The field curvature couples the two modes. The quantity k_o^2 can be either positive or negative; the first term in Eq. (34) comes from the plasma compression and the second one is the contribution of the pressure destabilizing term identified in Sect. 4.

As usual, these equations possess a nontrivial solution if their determinant is non zero, which yields the following dispersion for ω^2:

$$\omega^4 - \left[(V_A^2 + V_{SM}^2)k_{\parallel}^2 + V_A^2 k_o^2\right]\omega^2 + V_A^2 V_{SM}^2 k_{\parallel}^2(k_{\parallel}^2 - 2\beta^*\mathcal{K}_C\mathcal{K}_\rho) = 0. \tag{35}$$

First note that if both $B_z = 0$ (the so-called Z-pinch configurations) and $m = 0$, this equation is degenerate: one of the roots is $\omega^2 = 0$ and the other root is $\omega^2 = V_A^2 k_o^2$. Instability then requires that $k_o^2 < 0$, as $k_{\parallel} = 0$ in this case. This constrain is identical to the criterion[6] derived by Kadomtsev from the Energy Principle for the $m = 0$ mode in Z pinches (see Freidberg [12], p. 286).

When $m \neq 0$, Eq. (35) can be solved exactly but it is more instructive to analyze its properties. As the coefficient of ω^2 is equal to the sum of the two roots, and the last term is equal to their product, one finds that if $k_{\parallel}^2 > 2\beta^*\mathcal{K}_C\mathcal{K}_\rho$, the two roots are stable, and if $k_{\parallel}^2 < 2\beta^*\mathcal{K}_C\mathcal{K}_\rho$, one of the roots is unstable. If $B_z = 0$ (Z pinch), this condition is identical to the criterion[7] derived by Kadomtsev for $m \neq 0$ modes (see Freidberg [12] pp. 284–285).

---

[6] Kadomtsev's criterion for the $m = 0$ mode in a Z pinch is a necessary and sufficient condition of instability, whereas the analysis presented here shows only the sufficiency of this condition.

[7] Same comment as in the previous footnote.

Note that all these conditions of instability require $\mathcal{K}_C\mathcal{K}_P > 0$, in agreement with the discussion of Sects. 2 and 4; this condition is unavoidable in magnetically self-confined jets. The analysis presented here also shows that once this condition is satisfied, instability necessarily follows in static columns where $|B_\theta| \gg |B_z|$ on some of the radial range, as the magnetic tension stabilizing effect $V_A^2 k_\parallel^2$ is arbitrarily small in the vicinity of a magnetic resonance.

Finally, the reader may ask how the local analysis presented here informs us on the global stability properties of the column. The answer lies in the oscillation theorem of Goedbloed and Sakanaka [14]. The theorem states that for any (m, k) unstable mode, the growth rate decreases when increasing the number of radial nodes. This implies that if an unstable mode with a large number of radial nodes is found (such as the modes considered here), an unstable nodeless mode will also exist, and this mode will have the largest growth rate. Such a mode will have a very disruptive effect on the plasma if its displacement is not vanishing on the boundary, as will be the case if the azimuthal field is dominant on the boundary.

6 Moving Columns

The previous section has shown that cylindrical columns with a predominant azimuthal magnetic field at least in some radial range are subject to pressure-driven instabilities. This situation holds in the outer region of self-confined magnetic jets, leading to a potentially disruptive configuration. However, in these regions a gradient of axial velocity due to the interaction of the moving jet with the outside medium is also expected to be present, and it is legitimate to investigate the effect of such a velocity gradient on the stability properties of pressure-driven modes.

This problem has not yet been addressed in the astrophysics literature, but some relevant results are available in the fusion literature. In all the investigations cited below, the adopted velocity profile contains no inflexion point, in order to avoid the triggering of the Kelvin–Helmhotz instability.

It is first useful to consider what becomes of Suydam criterion in the presence of background motions.[8] This investigation has been performed by Bondeson et al. [6]. Focusing on axial flows ($\mathbf{U} = U_z(r)\mathbf{e}_r$), they conclude that the behavior of localized modes depends on the magnitude of

$$M \equiv \rho^{1/2}\frac{U_z'}{q'B_z/q},$$

(36)

where the prime denotes radial derivative, and q is the safety factor (see Sect. 3). This quantity is a form of Alfvénic Mach number based on the velocity

---

[8] This requires a generalization of Eq. (21); also, the Energy Principle no longer applies as the resulting operator is not self-adjoint.

and magnetic shear, hence its name. When $M^2 < \beta$, the flow shear destabilizes resonant modes. Above this limit, these modes are stable, but in this case, unstable modes exist at the edge of the slow continuum, and may be global. The authors found, however, that in this case the growth rates are small (comparable to the resistive instabilities growth rates). Note also that, as $q'/q \sim 1/r$, $M \sim (B_\phi/B_z)(r/d)(U_z/V_A) \gg 1$ in MHD jets (d is the width of the velocity layer).

These results seem to suggest that the region where the velocity shear layer takes place at the jet boundary is substantially stabilized in MHD jets. This seems to be confirmed by global linear stability analyses, both for interchange and ballooning modes, except possibly for the $m = 0$ ("saussage") mode [7] [23] [24]. In all cases, increase of the flow Mach number efficiently reduces the amplitude of the displacement of the unstable modes at the plasma boundary, an important feature to avoid the disruption of the plasma.

An efficient stabilization mechanism has also been identified in the nonlinear regime by Hassam [16]. This author exploits an analogy between the $m = 0$ pressure-driven interchange mode and the Rayleigh–Taylor instability in an appropriately chosen magnetized plasma configuration. From this analysis, he concludes that the $m = 0$ pressure-driven mode is nonlinearly stabilized by a smooth velocity shear $(dU_z/dr \sim U/R_o)$ if $M_s = U_z/C_S \gtrsim [\ln(\tau_d/\tau_g)]^{1/2}$, where τ_g is the instability growth time-scale $(\tau_g \sim c_s(\mathcal{K}_\rho\mathcal{K}_C)^{1/2})$ and τ_d the diffusion time-scale $(\tau_d \sim \nu\mathcal{K}_\rho\mathcal{K}_C$ where ν is the viscosity, assumed comparable to the resistivity). The nonlinear evolution of an unstable, slightly viscous, and resistive Z-pinch (i.e., a configuration where the field is purely azimuthal), was simulated by Desouza-Machado et al. [8]. They found that the plasma relaminarizes over almost all its volume for applied a velocity shear in good agreement with this analytic estimate. The core of the plasma still has some residual unstable "wobble", which can apparently be stabilized by the magnetic shear if some longitudinal field B_z is added to the configuration. Note that the large values of τ_d/τ_g relevant to astrophysical jets lead to only weak constraints[9] on the Mach number M_s, so that this nonlinear stabilization mechanism is expected to be efficient in astrophysical jets.

7 Summary and Open Issues

Pressure-driven instabilities occur in static columns when the pressure force pushes the plasma out from the inside of the magnetic field lines curvature, as shown from direct inspection of the "potential energy" of the linearized displacement equation (Sect. 4), and from a the dispersion relation of local modes (Sect. 5). When unstable modes exist, the growth rates are of the order C_S/R_o

---

[9] For example $\tau_d/\tau_g = 10^{30}$ translates into $M_s \gtrsim 8$ only; in YSO jets, this ratio is most probably significantly smaller, and the constraint is even weaker.

where C_S is the sound speed and R_o the jet radius. These are very large, comparable to the Kelvin–Helmholtz growth rate (the most studied instability in jets), especially that the ratio of the magnetic energy to the gas internal energy is expected to be of the order of unity (within an order of magnitude or so). Such instabilities are known to be disruptive in the fusion context when the eigenmodes exhibit a substantial displacement of the plasma outer boundary; such a situation is relevant to magnetically self-confined jets, as the magnetic field in their outer region is predominantly azimuthal, a configuration most favorable to the onset of the instability (Sects. 2 and 5). However, the presence of a velocity gradient in the outer boundary due to the jet bulk motion is expected to have a substantial stabilizing influence, both in the linear and nonlinear regimes (Sect. 6).

In its present state, this picture possesses a number of loose ends:

• The stabilizing role of an axial velocity gradient needs to be better understood. Not all modes may be stabilized in the linear regime, depending on the details of the equilibrium jet configuration, and the nonlinear mechanism identified in the literature is highly idealized and may not be generic. The one and only simulation of nonlinear stabilization published to date exhibits a very violent relaxation transient, which may still lead to jet disruption. On the other hand, this transient is also an indication that the initial configuration of the simulation is way out of equilibrium, a situation which may not occur in real jets.

• The role of jet rotation has not yet been correctly investigated. Preliminary results seem to indicate that it is stabilizing [17]; however, jet rotation may not be an important dynamical factor in the asymptotic jet propagation regime.

• Most investigations of pressure-driven instabilities rely on a very simple prescription of the equation of state, which raises an issue of principle. Indeed, the very large growth rates usually found for the instability indicate that it develops on time-scales much shorter than the collisional time-scale, and the use of ideal MHD as well as a polytropic equation of state may be questioned in such a context, an issue briefly addressed in Appendix A. A more complex description of the plasma is required to validate the results obtained so far.

Appendices

A On the Use of Ideal MHD

In astrophysics in general, and jet stability analyses in particular, an MHD framework is almost always adopted instead of the more precise kinetic one due to its relative simplicity. The MHD approximation can be applied when the fluid is locally neutral, when all species can be described by a single fluid equation (i.e., when the relative drift velocity of species with respect to one another is small), and when Ohm's law is valid. The validity of these approximations has been discussed elsewhere [18] [22] and will not be reproduced here; the interested reader is referred to these books for details.

Furthermore, MHD stability (and jet stability in particular) is often investigated within the framework of ideal MHD. Indeed, the dynamical time-scales of interest (including those of the considered instability) is almost always substantially larger than the particle collision time-scale. Moreover, an isotropic pressure is often assumed, e.g., through a barotropic equation of state, and this raises another issue of principle within the framework of ideal MHD, as, indeed, an isotropic pressure would be expected only if collisions at the particle level are not neglected.

The isotropic pressure assumption can be justified to some extent by the fact that plasma microturbulence does limit pressure anisotropy to a factor of order unity. For example, within the framework of collisionless MHD, pressure anisotropy is self-limiting (for a recent synthetic discussion of this problem within the framework of collisionless magnetorotational instability, see Sharma et al. [21] and references therein). Nevertheless, this provides little support (if any) to the adoption of a closure in the form of a barotropic or adiabatic equation of state in a collisionless setting.

The CGL approximation (or other collisionless MHD approximations) applies when the length-scales and frequencies under consideration are larger than the ion Larmor radius, and smaller than the ion cyclotron frequency, respectively. These conditions should be satisfied in jets, but I am not aware of any investigation of pressure-driven instabilities in this framework. Freidberg [12] argues that a simple rule of thumb to estimate the effects of the assumed closure is to replace the adiabatic index by 0, and to assume incompressibility of the motions within the framework of standard ideal MHD.

B Derivation of the Dispersion Relation

Some intermediate steps in the derivation of the dispersion relation of Sect. 5 are given here. The notations are the same as in this section.

Direct computation of the pressure and magnetic field perturbation gives

$$\delta P = -\rho_o C_S^2 \left[\boldsymbol{\nabla} \cdot \boldsymbol{\xi} + \mathcal{K}_\rho \xi_r \right], \tag{37}$$

$$\delta \boldsymbol{B} = -\boldsymbol{B}_o \left[\boldsymbol{\nabla} \cdot (\boldsymbol{\xi}_r + \boldsymbol{\xi}_l) + \xi_r (\mathcal{K}_B + \mathcal{K}_C) \right] \boldsymbol{e}_\parallel$$
$$- i k_\parallel B_o (\boldsymbol{\xi}_r + \boldsymbol{\xi}_l). \tag{38}$$

This allows us to obtain the perturbation in total pressure and in magnetic tension:

$$\delta P_T = -\rho_o (V_A^2 + C_S^2) \boldsymbol{\nabla} \cdot (\boldsymbol{\xi}_r + \boldsymbol{\xi}_l) + i \rho_o C_S^2 k_\parallel \xi_\parallel -$$
$$\rho_o \boldsymbol{\xi}_r \cdot (V_A^2 \mathcal{K}_B + V_A^2 \mathcal{K}_C + C_S^2 \mathcal{K}_\rho), \tag{39}$$

$$\delta T = -V_A^2 \mathcal{K}_C \left[2\boldsymbol{\nabla} \cdot (\boldsymbol{\xi}_r + \boldsymbol{\xi}_l) + 2\boldsymbol{\xi}_r \cdot (\mathcal{K}_B + \mathcal{K}_C) \right] +$$
$$ik_\| V_A^2 \left[\boldsymbol{\nabla} \cdot (\boldsymbol{\xi}_r + \boldsymbol{\xi}_l) + 2\boldsymbol{\xi}_r \cdot \mathcal{K}_C \right] e_\| - k_\|^2 V_A^2 (\boldsymbol{\xi}_r + \boldsymbol{\xi}_l). \tag{40}$$

Furthermore, the equilibrium relation Eq. (16) allows us to eliminate $\mathcal{K}_B$ in terms of $\mathcal{K}_C$ and $\mathcal{K}_\rho = \gamma \mathcal{K}_P$.

The longitudinal component of the linearized momentum equation reduces to $\delta P_T = 0$, once contributions of order $k_\| \xi_l$ or ξ_l / r are neglected in front of $k_\perp \xi_l$. This constraint shows that $\xi_l \sim O(k_\|/k_\perp \xi_r, k_\|/k_\perp \xi_\|)$. It also allows us to eliminate $\boldsymbol{\nabla} \cdot (\boldsymbol{\xi}_r + \boldsymbol{\xi}_l)$ from the remaining two component equations, which then reduce to Eqs. (32) and (33). In the process, the contribution of $d\delta P_T/dr$ is shown to be negligible from the assumed ordering relations $|k_r| \ll k_\perp$ and $|B_z/B_\theta|^2 s^2 |k_\perp| \ll |k_\||$, i.e., the magnetic shear stabilizing term can be neglected in this limit.

References

1. Begelman, M.C.: Instability of toroidal magnetic field in jets and plerions. **493**, 291 (1998)
2. Bernstein, I.B., Frieman, E.A., Kruskal, M.D., Kulsrud, R.M.: *Proc. Roy. Soc. London A* **244**, 17 (1958)
3. Blandford, R.D., Rees, M.J.: A 'twin-exhaust' model for double radio sources. Monthly Not. Royal Astron. Soc. **169**, 395 (1974)
4. Bodo, G., Massaglia, S., Rossi, P., Rosner, R., Malagoli, A., Ferrari, A.: The long-term evolution and mixing properties of high Mach number hydrodynamic jets. Astron. Astrophys. **303**, 281 (1995)
5. Bodo, G., Rossi, P., Massaglia, S., Ferrari, A., Malagoli, A., Rosner, R.: Three-dimensional simulations of jets. Astron. Astrophys. **333**, 1117 (1998)
6. Bondeson, A., Iacono, R., Bhattacharjee, A.: Local magnetohydrodynamic instabilities of cylindrical plasma with sheared equilibrium flows. Phys. Fluids **30**, 2167 (1987)
7. Chiueh, T.: Suppression of the edge interchange instability in a straight tokamak. Phys. Rev. E **54**, 5632–5635 (1996)
8. Desouza-Machado, S., Hassam, A.B., Sina, R.: Stabilization of Z pinch by velocity shear. Phys. Plasmas **7**, 4632–4643 (2000)
9. Dewar, R.L., Glasser, A.H.: Ballooning mode spectrum in general toroidal systems. Phys. Fluids **26**, 3038–3052 (1983)
10. Dewar, R.L., Tatsuno, T., Yoshida, Z., Nührenberg, C., McMillan, B.F.: Statistical characterization of the interchange-instability spectrum of a separable ideal-magnetohydrodynamic model system. Phys. Rev. E **70**, 066409 (2004)
11. Ferreira, J.: Magnetically-driven jets from Keplerian accretion discs. Astron. Astrophys. **319**, 340–359 (1997)
12. Freidberg, J.P.: Ideal Magnetohydrodynamics. Plenum Press, New York (1987)
13. Goedbloed, J.P.: Stabilization of magnetohydrodynamic instabilities by force-free magnetic fields. II. Linear pinch. Physica **53**, 501–534 (1971)
14. Goedbloed, J.P., Sakanaka, P.H.: Phys. Fluids **17**, 908 (1974)

15. Hain, K., Lüst, R.: Z. Naturforsch. Teil A **13**, 936 (1958)
16. Hassam, A.B.: Nonlinear stabilization of the Rayleigh-Taylor instability by external velocity shear. Phys. Fluids B **4**, 485–487 (1992)
17. Kersalé, E., Longaretti, P.Y., Pelletier, G.: Pressure- and magnetic shear-driven instabilities in rotating MHD jets. Astron. Astrophys. **363**, 1166–1176 (2000)
18. Krall, N.A., Trievelpiece, A.W.: Principles of Plasma Physics. San Francisco Press, San Francisco (1986)
19. Li, Z.: Magnetohydrodynamic disk-wind connection: magnetocentrifugal winds from ambipolar diffusion-dominated accretion disks. Astrophys. J. **465**, 855–+ (1996)
20. Ryu, D., Jones, T.W., Frank, A.: The magnetohydrodynamic Kelvin-Helmholtz instability: a three-dimensional study of nonlinear evolution. Astrophys. J. **545**, 475–493 (2000)
21. Sharma, P., Hammett, G.W., Quataert, E., Stone, J.M.: Shearing box simulations of the MRI in a collisionless plasma. ApJ**637**, 952 (2006)
22. Shu, F.H.: Physics of Astrophysics, Vol. II. University Science Books, Mill Valley (1992)
23. Shumlak, U., Hartman, C.W.: Sheared flow stabilization of the m = 1 Kink Mode in Z Pinches. Phy. Rev. Lett. **75**, 3285–3288 (1995)
24. Waelbroeck, F.L., Chen, L.: Ballooning instabilities in tokamaks with sheared toroidal flows. Phys. Fluids B **3**, 601–610 (1991)

Thermal Instabilities

G. Bodo

INAF, Osservatorio Astronomico di Torino, 10025 Pino Torinese, Italy,
`bodo@oato.inaf.it`

Abstract In this chapter I will discuss instabilities driven by a loss of energetic
equilibrium, both in the unmagnetized and in the magnetized case. I will first give
a derivation of the instability conditions by using a simple physical analysis and
then I will give a more formal derivation. I will finally briefly discuss the effects of
cooling-heating processes on the Kelvin-Helmholtz instability.

Keywords Hydrodynamics · interstellar medium: jets and outflows · thermal
instabilities · radiation losses

1 Introduction

The instabilities discussed in the previous chapter derived from a loss of mechanical equilibrium; those discussed in this chapter derived instead from a loss of energetic equilibrium. Their behavior is therefore dictated by the dependence of the processes of energy gains or losses on the physical parameters. In particular, in the case of jets, the most important process with which we have to deal is that of radiative losses. A first discussion of the possibility of thermal instabilities in an astrophysical plasma was presented by [8]. He considers, however, a rather nonphysical situation capable only of constant-density (isochoric) perturbations. Reference [11] recognized that gaseous systems tend to maintain constant pressure conditions and derived the correct isobaric instability criterion. A complete and fully consistent analysis, taking into account also the effect of several additional physical processes, was then given by [4]. In the next section, I will give the derivation of the instability condition starting from a simple physical analysis, a more formal derivation of the full dispersion relation will be presented in Sect. 3, where I will also discuss the dispersion relation for the magnetic case. Finally, in the last section, I will briefly discuss the effects of cooling–heating processes on the Kelvin–Helmholtz instability.

2 Physical Discussion

In an infinite, uniform medium, subject to energy gains and losses, the equilibrium condition can be written as

$$L(\rho_0, T_0) = 0$$

where ρ_0 and T_0 are, respectively, the density and temperature of the medium, while L is a generalized heat-loss function, defined as $L = \Lambda - H$, i.e., the difference between energy losses Λ and energy gains H per unit mass per unit time. Note that L can be written as a function of the local values of density and temperature if the medium is optically thin. If we introduce a perturbation of density $\delta\rho$ and a perturbation of temperature δT, such that the pressure is kept constant, we will have a corresponding variation of L given by:

$$\delta L = \frac{\partial L}{\partial T}\delta T + \frac{\partial L}{\partial \rho}\delta\rho \qquad (1)$$

The condition of isobaricity, assuming that the perfect gas law is applicable, can be written as

$$\frac{\delta T}{T_0} + \frac{\delta\rho}{\rho_0} = 0 \qquad (2)$$

Introducing Eq. (2) in Eq. (1), we can write

$$\frac{\delta L}{\Lambda_0} = (L_T - L_\rho)\frac{\delta T}{T} \qquad (3)$$

where $L_T = T_0/\Lambda_0 \partial L/\partial T$ and $L_\rho = \rho_0/\Lambda_0 \partial L/\partial \rho$. Here and after, the subscript 0 refers to the equilibrium values. If the perturbations of temperature and of L have opposite signs, the equilibrium is unstable since a decrease of temperature leads to an increase of losses and, therefore, to a runaway situation. The instability condition can be written, therefore, as

$$L_T - L_\rho < 0 \qquad (4)$$

For checking the validity of the assumption of isobaricity, we have to first define two typical time scales of the system: the dynamical time scale and the cooling (heating) time scale. The dynamical time scale τ_d can be defined as the, sound crossing time over the perturbation scalelength λ, i.e.,

$$\tau_d = \frac{\lambda}{c_s} \qquad (5)$$

where c_s is the sound speed. The cooling (heating) time, scale τ_r, for a perfect gas, can be defined as

$$\tau_r = \frac{p_0}{(\gamma - 1)\Lambda_0} \qquad (6)$$

where p_0 is the equilibrium pressure and γ is the ratio of specific heats. If the condition $\tau_r >> \tau_d$ holds, the medium has time to restore pressure equilibrium over the perturbation, that is, cooling (heating) with a time, scale τ_r.

Looking at the instability condition (4), we can see that a positive ρ-derivative increases the tendency towards instability. This condition is often met in astronomical applications, owing to the binary collision nature of radiative losses. The unstable mode is often called condensation mode since, in a cooling perturbation, the condition of isobaricity leads to an increase of density.

If we include a magnetic field, we have a different behavior of perturbations that are perpendicular to the field with respect to those that are not. The second involve motions along the field lines that are unimpeded by the presence of magnetic field and the instability criterion will be equal to the one derived above. For perturbations perpendicular to the field, we can derive a different instability criterion by following a similar line of thought. The isobaricity condition can be rewritten by using total (gas plus magnetic) pressure:

$$\beta \frac{\delta T}{T_0} + \beta \frac{\delta \rho}{\rho_0} + 2 \frac{\delta B}{B_0} = 0 \qquad (7)$$

where β is the ratio between thermal and magnetic pressure and B is the magnetic field intensity. In addition, the frozen-in condition gives a relation between density and magnetic field perturbations:

$$\frac{\delta B}{B} = \frac{\delta \rho}{\rho} \qquad (8)$$

The condition for instability can then be written as

$$L_T - \frac{\beta}{2 + \beta} L_\rho < 0 \qquad (9)$$

where we assumed that heat-loss mechanisms do not depend on magnetic field. There may however, be, situations in which there is a dependence of the function L on the magnetic field, for example, in the case of synchrotron losses (see e.g. [2]). We can note that, decreasing the value of β, the effect of the dependence of L on density tends to be less and less important and, in the limit $\beta \to 0$, the instability condition becomes the isochoric one [8]

$$L_T < 0 \qquad (10)$$

This behavior can be understood by considering that magnetic field opposes the plasma compression.

The instability criterion has been generalized to the case when the medium starts from a nonequilibrium situation by [1]. We consider a gas subject to heating and cooling given as before by the function L. In this case, however, we do not restrict ourselves to the case $L = 0$. The variation of the specific entropy in the time dt is given by

$$dS = \frac{dQ}{T} = -\frac{L}{T} dt \qquad (11)$$

Assume that a parcel of gas is slightly perturbed. We denote the difference in entropy between the parcel and its surrounding by δS, and similarly for all other variables. Consider local perturbations that satisfy the isobaric condition, then with $\delta p = 0$ we have

$$\frac{d}{dt} \delta S = \delta \frac{dS}{dt} = -\delta \frac{L}{T} \qquad (12)$$

If δS and $\delta(L/T)$ have different signs, the parcel entropy tends to evolve away from the background entropy and we have instability. Thus the instability criterion is

$$\left[\frac{\partial}{\partial S} \left(\frac{L}{T} \right) \right]_p < 0 \qquad (13)$$

that can be written in the form (for $L_0 \neq 0$)

$$L_T - L_\rho < 1 \qquad (14)$$

that can be directly compared with Eq. (4) above.

3 Linear Analysis

In this section, I will give a formal derivation of the dispersion relation for the thermal instability in a uniform medium, including the effect of thermal conduction. We start from the nonmagnetic case, and the basic equations are

$$\frac{d\rho}{dt} + \rho\nabla \cdot \mathbf{v} = 0 \tag{15}$$

$$\rho\frac{d\mathbf{v}}{dt} + \nabla p = 0 \tag{16}$$

$$\frac{1}{\gamma - 1}\frac{dp}{dt} - \frac{\gamma}{\gamma - 1}\frac{p}{\rho}\frac{d\rho}{dt} + \rho L - \nabla \cdot (K\nabla T) = 0 \tag{17}$$

where d/dt is $\partial/\partial t + \mathbf{v} \cdot \nabla$, p, ρ and T are, respectively, pressure, density and temperature, γ is the ratio of specific heats, and K is the coefficient of thermal conduction. System (15, 16, 17) has to be complemented by the equation of state for perfect gases. The equilibrium configuration is characterized by $\rho = \rho_0$, $T = T_0$, $\mathbf{v} = 0$, and $L(\rho_0, T_0) = 0$. Assuming perturbations of the form

$$f(\mathbf{r}, t) = \tilde{f}_1 \exp(\tilde{\sigma}t + i\tilde{\mathbf{k}} \cdot \mathbf{r}) \tag{18}$$

the linearized equations take the form

$$\tilde{\sigma}\tilde{\rho}_1 + \rho_0 i\tilde{\mathbf{k}} \cdot \tilde{\mathbf{v}}_1 = 0 \tag{19}$$

$$\tilde{\sigma}\rho_0\tilde{\mathbf{v}}_1 + i\mathbf{k}\tilde{p}_1 = 0 \tag{20}$$

$$\frac{\tilde{\sigma}}{\gamma - 1}\tilde{p}_1 - \frac{\tilde{\sigma}\gamma p_0}{(\gamma - 1)\rho_0}\tilde{\rho}_1 + \rho_0\frac{\partial L}{\partial \rho}\tilde{\rho}_1 + \rho_0\frac{\partial L}{\partial T}\tilde{T}_1 + K_0 k^2\tilde{T}_1 = 0 \tag{21}$$

$$\frac{\tilde{p}_1}{p_0} - \frac{\tilde{\rho}_1}{\rho_0} - \frac{\tilde{T}_1}{T_0} = 0 \tag{22}$$

We can nondimensionalize these equations by measuring density, pressure, and temperature with their equilibrium values, velocity with the sound speed, time with the radiative time scale, and length with the length ($\lambda_r = c_s\tau_r$) crossed in a radiative time at the sound speed. We introduce, therefore, the following nondimensional quantities

$$\rho_1 = \frac{\tilde{\rho}_1}{\rho_0}, \quad p_1 = \frac{\tilde{p}_1}{p_0}, \quad T_1 = \frac{\tilde{T}_1}{T_0}, \quad \mathbf{v}_1 = \frac{\tilde{\mathbf{v}}_1}{c_s}$$

$$\sigma = \tilde{\sigma}\tau_r, \quad \mathbf{k} = \tilde{\mathbf{k}}\tau_r c_s, \quad \alpha = \frac{(\gamma - 1)}{p_0}\frac{K_0 T_0}{c_s^2\tau_r}$$

where α measures the effect of thermal conduction and is the ratio between the mean free path of conducting particles and the radiative length λ_r. We can then rewrite Eqs. (19, 20, 21, 22) in the form

$$\sigma\rho_1 + i\mathbf{k} \cdot \mathbf{v}_1 = 0 \tag{23}$$

$$\sigma\mathbf{v}_1 + i\frac{1}{\gamma}\mathbf{k}p_1 = 0 \tag{24}$$

$$\sigma p_1 - \gamma\sigma\rho_1 + (L_T - L_\rho) + \alpha k^2 T_1 = 0 \tag{25}$$

$$p_1 - \rho_1 - T_1 = 0 \tag{26}$$

Since $\mathbf{k} \times \mathbf{v_1} = 0$, we are left with four equations in the four variables ρ_1, p_1, T_1, and $\mathbf{k} \cdot \mathbf{v_1}$, and the dispersion relation can be obtained imposing the vanishing of the determinant of coefficients of the above system of equations. We then derive the following dispersion relation

$$\sigma^3 + \sigma^2 \left(L_T + \alpha k^2\right) + \frac{k^2}{k_r^2}\sigma + \frac{1}{\gamma}\frac{k^2}{k_r^2}\left[(L_T - L_\rho) + \alpha k^2\right] = 0 \qquad (27)$$

The three roots of the above equation correspond to three modes, one is the condensation mode, whose physics has been described in Sect. 2, and the other two correspond to sound waves modified by nonadiabaticity. If cooling–heating effects are small one root σ_1 is real (condensation mode) and the other two σ_2 and σ_2^* are complex conjugates (sound waves, for a proof see [4]), in this case

$$\sigma_1 |\sigma_2|^2 = -\frac{1}{\gamma}\frac{k^2}{k_r^2}\left[(L_T - L_\rho) + \alpha k^2\right] \qquad (28)$$

The condition for instability $\sigma_1 > 0$ then becomes

$$L_T - L_\rho + \alpha k^2 < 0 \qquad (29)$$

that corresponds to the isobaric condition 4 modified by the presence of the stabilizing effect of thermal conduction. In Fig. 1, we plot the growth rate of the condensation mode as a function of the wavenumber k for different values of the thermal conduction parameter α.

In the case with no thermal conduction (solid curve), we can see that the growth rate, in the limit of large wavenumbers, becomes independent from the wavenumber. In fact, the isobaric condition becomes better and better verified. Decreasing the wavenumber, the cooling time becomes comparable to the dynamical time and pressure equilibrium cannot be re established. Therefore, as the wavenumber tends to zero, we are approaching an isochoric situation and the behavior of the growth rate depends on whether the isochoric instability criterion is satisfied or not. In the case presented in Fig. 1, the parameters are such that the isochoric criterion is not satisfied and the growth rate tends to zero as $k \to 0$. On the contrary, when the isochoric instability criterion is satisfied, the growth rate tends to a finite value as $k \to 0$. The introduction of thermal conduction, as it is shown in the figure, stabilizes large wavenumbers and implies the existence of a wavenumber at which the growth rate attains a maximum. This maximum defines the scale of perturbations that will grow fast and will dominate the evolution.

The introduction of a magnetic field modifies the basic equations introducing the magnetic force term in the momentum equation and the induction equation for the magnetic field evolution. In addition, thermal conduction becomes anisotropic with different conductivities along and across the field lines. The dispersion relation takes the following form

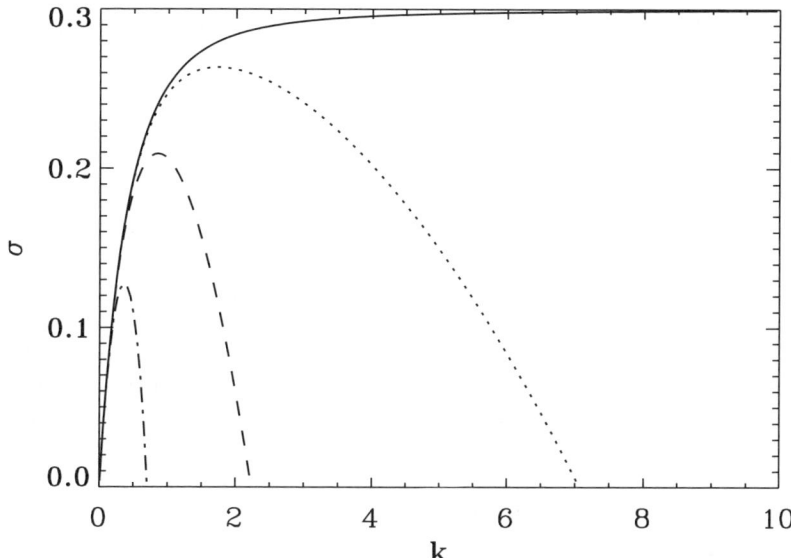

Fig. 1. Plot of the growth rate σ as a function of the wavenumber k. The four curves are for four different values of the thermal conduction parameter α ($\alpha = 0$ *solid curve*, $\alpha = 0.01$ *dotted curve*, $\alpha = 0.1$ *dashed curve*, $\alpha = 1$ *dashed-dotted curve*). The parameters are $L_T = 0.5$, $L_\rho = 1$, $\gamma = 5/3$

$$\sigma^5 + \frac{k^2}{k_r^2}\left(L_T + \alpha_\| k_\|^2 + \alpha_\perp k_\perp^2\right)\sigma^4 + \frac{k^2}{k_r^2}\left(1 + \frac{2}{\gamma\beta}\right)\sigma^3 +$$

$$+ \frac{1}{\gamma}\frac{k^2}{k_r^2}\left[L_T - L_\rho + \alpha_\| k_\|^2 + \alpha_\perp k_\perp^2 + \frac{2}{\beta}\left(L_T + \alpha_\| k_\|^2 + \alpha_\perp k_\perp^2\right)\right]\sigma^2 +$$

$$\frac{k^2}{k_r^2}\frac{k_\|^2}{k_r^2}\frac{2}{\gamma\beta}\sigma + \frac{k^2}{k_r^2}\frac{k_\|^2}{k_r^2}\frac{2}{\gamma\beta}\frac{1}{\gamma}\left(L_T - L_\rho + \alpha_\| k_\|^2 + \alpha_\perp k_\perp^2\right) = 0 \quad (30)$$

where

$$\alpha_\| = \frac{(\gamma - 1)}{p_0}\frac{K_{0\|}T_0}{c_s^2 \tau_r}, \qquad \alpha_\perp = \frac{(\gamma - 1)}{p_0}\frac{K_{0\perp}T_0}{c_s^2 \tau_r},$$

$K_{0\|}$ and $K_{0\perp}$ are the conductivities, respectively, along and across magnetic field lines and $k_\|$ and $k_\perp$ are the components of the wavevector, respectively, along and across magnetic field lines.

The dispersion relation is of fifth-order and its solution represent two modified fast modes, two modified slow modes, and the condensation mode. We can find approximate solutions in the isobaric limit $k^2/k_r^2 \ll 1$. With the additional assumption $k_\|^2/k_r^2 \ll 1$, we obtain for the growth rate of the condensation mode the following expression:

$$\sigma = -\frac{1}{\gamma}\left(L_T - L_\rho + \alpha_\| k_\|^2 + \alpha_\perp k_\perp^2\right) \quad (31)$$

that is equal to the isobaric growth rate of the condensation mode in the case without magnetic field, with the only modification of the anisotropic thermal conductivity. When the angle between the wavevector and magnetic field approaches $\pi/2$, the condition $k_\parallel^2/k_r^2 \ll 1$ becomes invalid and the behavior of the growth rate depends on whether the instability criterion for perpendicular wavevector is satisfied or not. If it is not satisfied the growth rate tend to zero; in the other case it tends to the value

$$\sigma = \frac{1}{\gamma}\left[L_T - L_\rho + \alpha_\parallel k_\parallel^2 + \alpha_\perp k_\perp^2 + \frac{2}{\beta}\left(L_T + \alpha_\parallel k_\parallel^2 + \alpha_\perp k_\perp^2\right)\right]$$

In Figs. 2 and 3, we show the behavior of the growth rate as a function of the parallel component of the wavenumber, keeping the fixed total wavenumber $k = 4$, respectively, for the cases without and with thermal conduction. The different curves correspond to different values of L_T going from $L_T = -2$ (solid curve) to $L_T = 0.5$ (dash-dotted curve). In the case without thermal conduction, we can see that, as $k_\parallel/k \to 1$, the growth rate tends to the value given by Eq. (31). For $k_\parallel \to 0$, we have instead the different behaviors described above. Thermal conduction, being anisotropic and larger in the parallel direction, introduces a stabilization of large parallel wavenumbers only. The maximum growth rate is then found for structures that are elongated in the field direction.

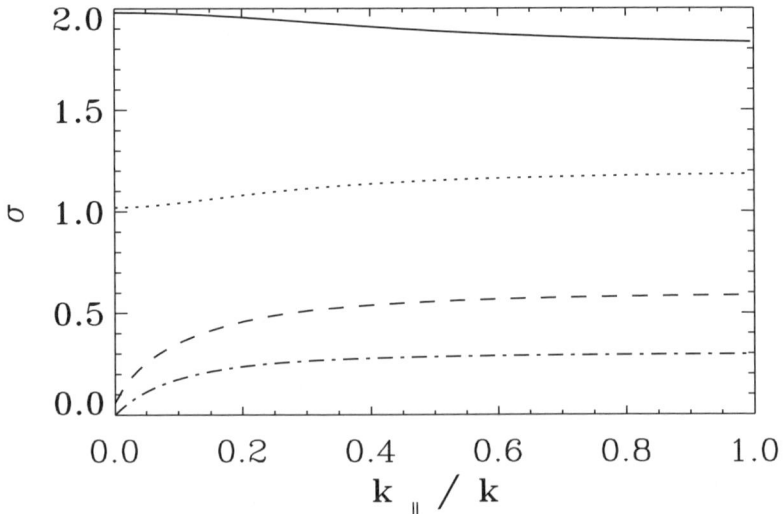

Fig. 2. Plot of the growth rate for the magnetized case versus the component of wavenumber parallel to magnetic field. The parameters are $k = 4$, $\beta = 0.067$, $L_\rho = 1$, no thermal conduction. The different curves refer to different values of L_T ($L_T = -2$ *solid*, $L_T = -1$ *dotted*, $L_T = 0$ *dashed*, $L_T = 0.5$ *dash-dotted*)

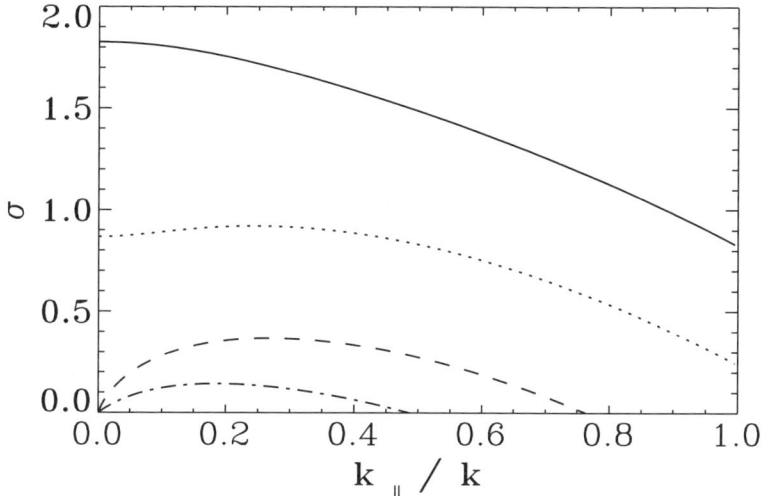

Fig. 3. Plot of the growth rate for the magnetized case versus the component of wavenumber parallel to magnetic field. The parameters are $k = 4$, $\beta = 0.067$, $L_\rho = 1$, $\alpha_\| = 0.1$, and $\alpha_\perp = 0.01$. The different curves refer to different values of L_T ($L_T = -2$ *solid*, $L_T = -1$ *dotted*, $L_T = 0$ *dashed*, $L_T = 0.5$ *dash-dotted*)

I conclude this analysis with a short consideration on the evolution of thermal instabilities. Typically it may lead to a two-phase situation, in which cool condensations are embedded into a hot and low-density medium. This structure can be reached if the temperature and density in the two phases correspond to stable energetic equilibria.

4 Influence of Radiative Losses on the KH Instability

Besides driving possible instabilities, the cooling–heating effects considered above may also influence the behavior of Kelvin–Helmholtz instabilities. This aspect, in the context of jets from YSO, has been considered by [5, 6] in the linear case and by [3, 7, 9, 10] for the nonlinear evolution. Reference [6] find that reflected modes are the most influenced by the presence of cooling effects that tend to stabilize them. The growth rates differ from the adiabatic case starting from $\tau_r/\tau_d < 10$. Moreover, according to their analysis, thermal modes, when they are unstable, may have growth rates larger than KH modes. The stabilizing behavior found by [6] may, however, be dependent on the form of radiative losses considered, i.e., on its variation with temperature. Different dependences may lead to different behaviors that could also be destabilizing [5]. This dependence on the form of radiative losses is present also in numerical simulations of the nonlinear evolution. However, we may say that, in general radiation increases the time taken for shocks to develop, reduces the strength of these shocks, and reduces the rate of decollimation of the momentum flux.

References

1. Balbus, S.A. ApJ, **303L**, 79 (1986)
2. Bodo, G., Ferrari, A., Massaglia, S., Trussoni, E. MNRAS, **244**, 530 (1990)
3. Downes, T.P., Ray, T.P. A&A, **331**, 1130 (1998)
4. Field, G. ApJ, **142**, 531 (1965)
5. Hardee, P.E., Stone, J.M. ApJ, **483**, 121 (1997)
6. Massaglia, S., Trussoni E., Bodo, G., Rossi, P., Ferrari, A. A&A, **260**, 243 (1992)
7. Micono, M., Bodo, G., Massaglia, S., Rossi, P., Ferrari, A., Rosner, R. A&A, **364**, 318 (2000)
8. Parker, E. ApJ, **117**, 431 (1953)
9. Rossi, P., Bodo, G., Massaglia, S., Ferrari, A. A&A, **321**, 672 (1997)
10. Stone, J.M., Xu, J., Hardee, P.E. ApJ, **483**, 121 (1997)
11. Weymann, R. ApJ, **132**, 452 (1960)

The Oscillatory Instability of Radiative Shock Waves

A. Mignone

Dipartimento di Fisica Generale dell'Università, Via Pietro Giuria 1, I-10125
Torino, Italy; INAF Osservatorio Astronomico di Torino, 10025 Pino Torinese,
Italy,
mignone@to.astro.it

Abstract. Shock waves undergoing radiative cooling are known to suffer from a
global instability caused by rapid variations of the cooling time scale with the shock
speed. In the limit of optically thin plasma, the linear stability analysis of planar
shocks with a power-law cooling function $\Lambda \propto \rho^2 T^\alpha$ is reviewed. It is shown that
the shock front can oscillate around its stationary position with increasing quantized frequencies resembling the modes of a pipe open at one end. Transition to
nonlinearity is investigated by solving the full time-dependent problem by means
of numerical simulations. For values of the sufficiently small ($\alpha \lesssim 0.5$), cooling exponent computations reveal that the shock oscillations are shown to saturate at a
finite amplitude and tend to a quasi periodic cycle.

Mignone, A.: *The Oscillatory Instability of Radiative Shock Waves.* Lect. Notes Phys. **754**,
163–176 (2008)
DOI 10.1007/978-3-540-76967-5_6   © Springer-Verlag Berlin Heidelberg 2008

Keywords Hydrodynamics · Instabilities · Methods: numerical · Shock waves · Stars · binaries · close

1 Introduction

In a radiative shock wave, the cooling time scale due to cooling processes becomes comparable to or less than the sound crossing time behind the wave front. Under these circumstances, the postshock flow may be subject to a global thermal instability (more precisely, an overstability given to the complex nature of the eigenvalues) revealed by rapid variations of the cooling time scale with the shock speed. The instability drives the shock front to oscillate with respect to its stationary position, causing fluctuations in the amount of radiation emitted from the postshock region.

Theoretical investigations of the instability date back to the early work of [4, 7, 18] and has been fostered by its pertinence to a number of different astrophysical environments including magnetic cataclysmic variables [6, 31], jets from young stellar objects [8], magnetospheric accretion in T-Tauri stars [5], colliding stellar winds [1, 23], and supernova remnants [3, 15, 27]. Similarly, a relevant issue arises in questioning the validity of steady shock models with shock velocities $v_s \gtrsim 130$ km s^{-1}, routinely used in interpreting emission line observations from interstellar shocks [13, 14].

In an old paper, [4] (CI hereafter) presented the linear stability analysis of planar radiative shocks with volumetric cooling rate given by a cooling function in the form $\Lambda \propto \rho^2 T^\alpha$. CI showed that the shock has multiple modes of oscillation with increasing frequency and the stability of a given mode is tied to the value of the cooling exponent α. Higher values of α were shown to stabilize the shock, whereas lower values of α promote the growth of instability. In this respect, the fundamental mode becomes unstable for $\alpha \lesssim 0.4$, while the first and second harmonics are destabilized when $\alpha \lesssim 0.8$. Higher order harmonics were not considered by CI.

Since then, the issue of stability has been investigated under a variety of different regimes: cyclotron emission [11, 29, 30], spherical geometry [2], gravitational effects [9], magnetic fields [26], two temperature flows [12], and multiple cooling functions [21, 22]. Additional references can be found in the review by [31].

Moreover, considerable efforts have been devoted to the solution of the full time-dependent problem by mean of numerical simulations. Indeed, the existence of the instability was first revealed in the numerical computations of [7, 16]. Additional work may be found [10, 17, 19, 24, 25, 28] and references therein. In [10], for instance, the critical values of α (above which oscillations are damped) were shown to lie somewhere in the range $1/3 \lesssim \alpha \lesssim 1/2$ and $1/2 \lesssim \alpha \lesssim 0.6$ for the fundamental and first overtone, respectively. The one-dimensional calculations of [24] showed that for flows incident into a wall, large amplitude oscillations are damped when $\alpha \gtrsim 0.5$. These results were recently extended by [25] to a more realistic cooling function. A quantitative comparison between theoretical predictions and numerical models can be found in [19], where it is shown that most overtones found in the simulation can be positively matched with the ones given by linear analysis, provided suitable boundary conditions are given.

The paper is organized as follows. In Sect. 2, the relevant equations are introduced and the stability properties of 1-D planar radiative shocks are reviewed. In Sect. 3, the transition to the nonlinear regime using numerical simulations is presented. Finally, in Sect. 4, we discuss some potential implications for a variety of astrophysical scenario.

2 Linear Theory

2.1 Statement of the Problem

Consider a one-dimensional supersonic flow with uniform density ρ_{in} and velocity v_{in}, initially propagating in the negative x-direction, i.e., $v_{\mathrm{in}} = -|v_{\mathrm{in}}|$. The flow is brought to rest by the presence of a rigid wall located at $x = 0$, and a shock wave forms at some finite distance x_{s} from the wall (see Fig. 1). Across the shock, the bulk kinetic energy of the incoming gas is converted into

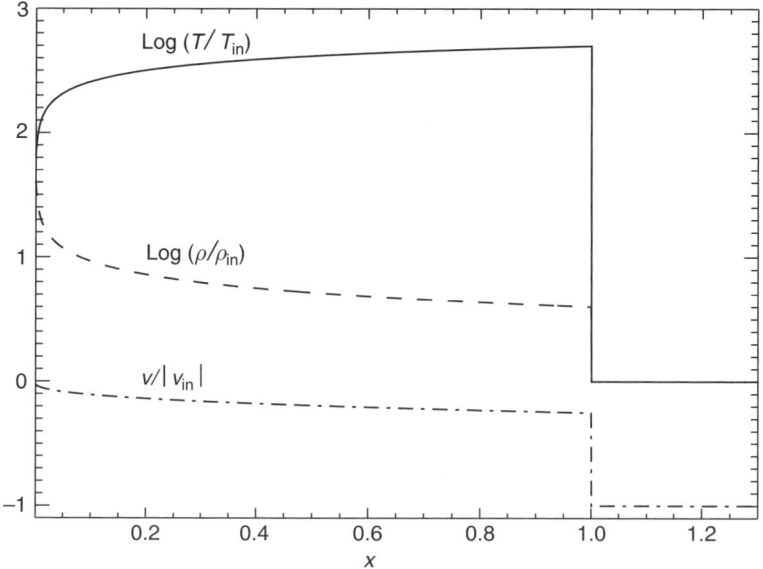

Fig. 1. Steady-state profiles for density, temperature, and velocity when $\alpha = 0$. The "wall" is located at $x = 0$ and supersonic gas flows from the right to the left. Flow variables are normalized to their inflow values, and the abscissa is expressed in units of shock height

thermal motion and the flow decelerates to subsonic velocities. Throughout the postshock region, the excess thermal energy is radiated away by cooling processes which, for our purposes, are assumed optically thin. In this approximation and to make the problem tractable, we make the further assumption that thermal losses can be described by a simple power-law cooling function $\Lambda(\rho, p) \propto \rho^2 T^\alpha$ where α is a free parameter of the problem. In its simplest form, the problem can be treated using the Euler equations of gas-dynamics:

$$\frac{\partial \rho}{\partial t} + \rho \frac{\partial v}{\partial x} + v \frac{\partial \rho}{\partial x} = 0, \tag{1}$$

$$\frac{\partial v}{\partial t} + v \frac{\partial v}{\partial x} + \frac{1}{\rho} \frac{\partial p}{\partial x} = 0, \tag{2}$$

$$\frac{\partial p}{\partial t} + v \frac{\partial p}{\partial x} + \gamma p \frac{\partial v}{\partial x} = -(\gamma - 1)\mathcal{C}\rho^2 \left(\frac{p}{\rho}\right)^\alpha, \tag{3}$$

where ρ and p are, respectively, the fluid density and pressure and $\gamma = 5/3$ is the constant specific heat ratio. By proper renormalization of the constant $\mathcal{C}$, one can express density and velocity in units of their inflow values, i.e., ρ_{in} and $|v_{\text{in}}|$. With this choice, the flow variables immediately ahead of the shock become $\rho = 1$, $v = -1$, and $p = 1/(\gamma \mathcal{M}^2)$, with $\mathcal{M}$ being the upstream Mach number. Furthermore, lengths are conveniently normalized to the stationary thickness of the cooling region, so that the equilibrium position of the shock is $x = 1$.

The steady state condition is dictated by an exact balance between dynamical and cooling time scales. In other words, a fluid element travels through the postshock region and radiates all its internal energy by the time it reaches the wall, where the temperature drops to zero. In equilibrium, flow quantities immediately ahead and behind the shock (denoted with the subscript s) satisfy the Rankine-Hugoniot jump conditions:

$$-v_s = \frac{1}{\rho_s} = \frac{\gamma - 1}{\gamma + 1} + \frac{2}{(\gamma + 1)\mathcal{M}^2}, \quad p_s = \frac{2}{\gamma + 1} - \frac{\gamma - 1}{\gamma(\gamma + 1)\mathcal{M}^2}, \tag{4}$$

If, however, the postshock temperature is slightly increased (for example), a longer cooling path will result and the excess pressure will force the shock to move away from the wall. In the frame of the shock, the velocity of the incoming gas will increase even further and the postshock temperature will rise according to the square of the preshock velocity. If radiative losses are described by a decreasing function of the temperature, the cooling time will increase and the shock will continue to move away from the wall.

2.2 Perturbative Analysis

A perturbative study is carried out by properly linearizing Eqs. (1, 2, 3) around the steady-state solutions denoted by ρ_0, v_0, and p_0. The perturbed location of the shock front is written as

$$x_\mathrm{s} = 1 + \frac{\epsilon}{\delta} e^{\delta t} , \qquad (5)$$

where $x_\mathrm{s} = 1$ is the shock equilibrium position, ϵ is the magnitude of the perturbation, and δ is a complex eigenfrequency to be determined.

Following [21], it is convenient to write $\delta = \delta_\mathrm{R} + i\delta_\mathrm{I}$, where the real part δ_R gives the growth/decay term, while δ_I represents the oscillation frequency. The nature of the instability is determined by the sign of δ_R: modes with $\delta_\mathrm{R} < 0$ are stable, while modes with $\delta_\mathrm{R} > 0$ are unstable.

We write a generic flow quantity as

$$q(\xi, t) = q_0(\xi) \left(1 + \lambda_q(\xi) \epsilon e^{\delta t} \right) , \qquad (6)$$

where $q \in \{\rho, v, p\}$, $q_0(\xi)$ is the corresponding steady-state value and the complex function $\lambda_q(\xi)$ describes the effects of the perturbation. Here ξ is a spatial coordinate normalized so that $\xi = 1$ at the shock and $\xi = 0$ at the wall:

$$\xi = \frac{x}{x_\mathrm{s}} \approx x \left(1 - \frac{\epsilon}{\delta} e^{\delta t} \right) + O(\epsilon^2) . \qquad (7)$$

The fluid equations are linearized in a frame of reference which is co-moving with the shock; in this frame the derivatives of a flow variable become

$$\frac{\partial}{\partial t} \rightarrow \frac{\partial}{\partial t} + \frac{\partial \xi}{\partial t} \frac{\partial}{\partial \xi} , \qquad \frac{\partial}{\partial x} \rightarrow \frac{\partial \xi}{\partial x} \frac{\partial}{\partial \xi} . \qquad (8)$$

Therefore, retaining only terms up to first-order in ϵ, one has

$$\frac{\partial q}{\partial t} \approx \left(q_0 \lambda_q \delta - \xi q_0' \right) \epsilon e^{\delta t} , \qquad (9)$$

$$\frac{\partial q}{\partial x} \approx q_0' + \left(q_0' \lambda_q + q_0 \lambda_q' - \frac{q_0'}{\delta} \right) \epsilon e^{\delta t} , \qquad (10)$$

where a primed quantity denotes a derivative with respect to ξ.

The steady-state solution is recovered by collecting the zeroth order terms in the Euler equations; conservation of mass and momentum is trivially expressed by

$$\rho_0 v_0 = -1 , \quad -v_0 + p_0 = m , \qquad (11)$$

where the integration constants on the right hand sides may be evaluated from the preshock values; hence $m = 1 + 1/(\gamma \mathcal{M}^2)$. The pressure equation (3) with $\partial p / \partial t = 0$ provides the explicit dependence on the spatial coordinate. After expressing ρ_0 and p_0 as functions of v_0 alone using Eq. (11), one finds the closed integral form

$$\xi(v_0) = \frac{f(v_0)}{f(v_\mathrm{s})} , \quad \text{with} \quad f(v) = \int_0^v (-y)^{2-\alpha} \frac{[y + \gamma(m + y)]}{(m + y)^\alpha} \, dy , \qquad (12)$$

and $v_\mathrm{s} = -(1 + 3/\mathcal{M}^2)/4$ is the fluid velocity immediately behind the shock (Eq. (4)). According to the normalization units introduced in the previous subsection, the constant $\mathcal{C}$ in Eq. (3) takes the value $\mathcal{C} = -f(v_\mathrm{s})/(\gamma - 1)$.

Results pertinent to this section are evaluated in the strong shock limit, $\mathcal{M} \to \infty$, so $m = 1$, $v_{\rm s} = -1/4$ and α becomes the only free parameter in the problem.

Although Eq. (12) can be solved analytically for some specific values of the cooling index α (CI), one has to resort to numerical quadrature in the general case. Note that a steady-state solution is possible only if the integral converges, that is, if $\alpha < 3$. For higher values of α, the gas cannot cool to zero temperature in a finite time. Equation (12) can be inverted numerically to express the postshock steady flow velocity v_0 as a function of ξ. The resulting steady-state profiles are shown in Fig. 1.

The first-order terms in ϵ provide three coupled complex differential equations for the perturbations; using the unperturbed postshock velocity v_0 as the independent variable, they are

$$\frac{d\lambda_\rho}{dv_0} + \frac{d\lambda_v}{dv_0} = -\frac{\xi}{v_0^2} - \frac{\lambda_\rho \delta}{v_0} \frac{d\xi}{dv_0}, \tag{13}$$

$$v_0 \frac{d\lambda_v}{dv_0} - p_0 \frac{d\lambda_p}{dv_0} = -\lambda_v \delta \frac{d\xi}{dv_0} + \frac{\xi}{v_0} + \lambda_p - 2\lambda_v - \lambda_\rho, \tag{14}$$

$$v_0 p_0 \left(\gamma \frac{d\lambda_v}{dv_0} + \frac{d\lambda_p}{dv_0} \right) = (v_0 + \gamma p_0) Q - p_0 \lambda_p \delta \frac{d\xi}{dv_0} + \xi, \tag{15}$$

where $d\xi/dv_0$ is given by straightforward differentiation of Eq. (12) and $Q = (2 - \alpha)\lambda_\rho + (\alpha - 1)\lambda_p - \lambda_v + 1/\delta$.

For a given value of α, Eqs. (13) through (15) have to be solved by integrating from the shock front (where $v_0 = v_{\rm s}$) to the wall (where $v_0 = 0$). The eigenmodes of the system are determined by imposing appropriate boundary conditions to select the physically relevant solutions. At the shock front, the jump conditions for a strong shock ($\mathcal{M} \to \infty$) apply [12, 21]:

$$\lambda_\rho = 0, \quad \lambda_v = -3, \quad \lambda_p = 2. \tag{16}$$

At the bottom of the postshock region ($\xi = 0$), the relevant physical solutions must satisfy the stationary wall condition, namely, that the flow comes to rest and the velocity must be oscillation-free. This requires that both the real and imaginary parts of $\lambda_v(v_0)$ vanish at $v_0 = 0$. The complex frequencies δ for which such solutions possibly identify the eigenmodes of the system.

In practice, one needs to minimize the real function of two variables $|\lambda_v(0)|(\delta_{\rm R}, \delta_{\rm I}) = \left(\lambda_{v,{\rm R}}^2(0) + \lambda_{v,{\rm I}}^2(0)\right)^{1/2}$, that is, the magnitude of the velocity perturbation at the bottom of the postshock region. The values of $\lambda_{v,{\rm R}}(0)$ and $\lambda_{v,{\rm I}}(0)$ are obtained by direct numerical integration of Eqs. (13, 14, 15) for a given pair $(\delta_{\rm R}, \delta_{\rm I})$, see Fig. 2. A preliminary coarse search with trial values of $\delta_{\rm R}$ and $\delta_{\rm I}$ reveals that, for a given value of α, an indefinitely long series of modes exists. Following CI, modes are labeled by increasing oscillation frequency, so that $n = 0$ corresponds to the fundamental mode, $n = 1$ to the

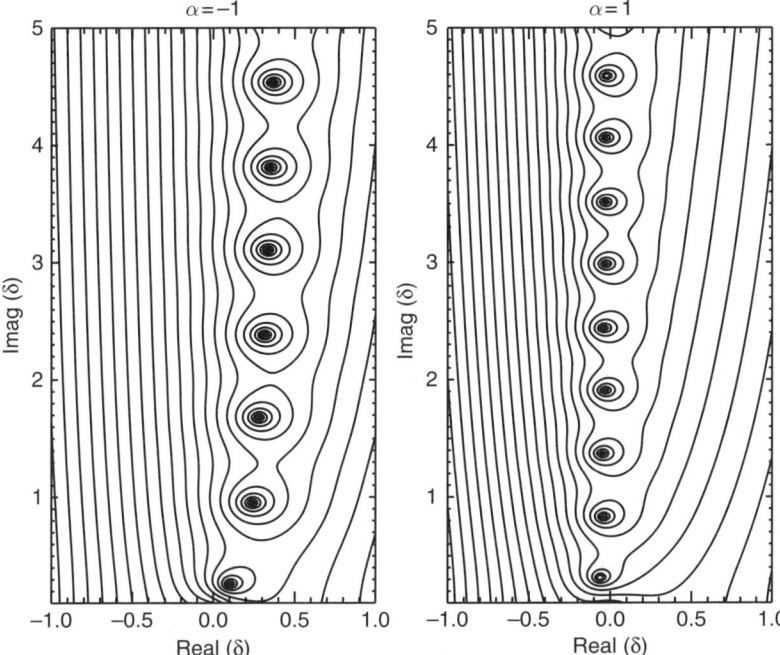

Fig. 2. Contour of $\log |\lambda_v|$ at $\xi = 0$ in the complex eigenplane $\delta = \delta_R + i\delta_I$ for $\alpha = -1$ (*left*) and $\alpha = 1$ (*right*). Physical admissible modes are located where $|\lambda_v| \to 0$

first overtone, $n = 2$ to the second overtone, and so on. The exact position of each mode n, $(\delta_R^{(n)}, \delta_I^{(n)})$, can be iteratively improved by repeating the search on finer subgrids (in the complex δ plane) centered around the most recent iteration of $\delta_R^{(n)}$, $\delta_I^{(n)}$.

It is important to realize that the physical admissible modes have been identified by imposing the stationary wall condition at $\xi = 0$. This constraint yields quantized modes of oscillation distinguished by their frequencies and the number of nodes and antinodes appearing in the appropriately scaled postshock structure, [32]. This behavior is thus similar to a standing vibrating string (or open pipe) fixed at one end. A typical astrophysical scenario corresponds to a standing accreting shock front formed above the surface of a white dwarf. At the stellar surface, the shocked flow is brought to rest (the wall at the closed end) whereas the shock can freely move (the open end).

Figure 3 shows the real and imaginary parts of the first eight modes for values of α uniformly distributed in the range $-2 \le \alpha < 2$.

A mode is stable if the real part of the corresponding eigenvalue has negative sign, and unstable otherwise. High-frequency modes are characterized by growth rates which decrease faster than low-frequency ones for increasing α. Hence, the fundamental mode ($n = 0$) has the smallest growth/damping rate

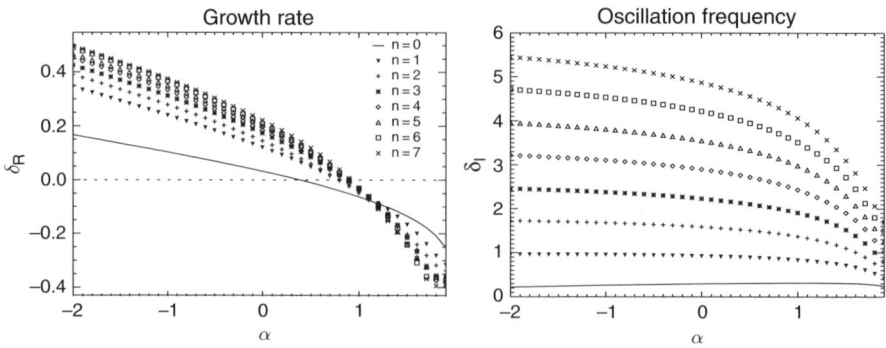

Fig. 3. Growth rates (*left*) and oscillation frequencies (*right*) for the first eight modes as function of α. The *solid line* represents the fundamental mode $n = 0$, whereas the different symbols (described by the legend in the *upper-right* portion of the plot) correspond to the seven overtones $1 \leq n \leq 7$. Eigenmodes with $\delta_R < 0$ are stable, whereas modes with $\delta_R > 0$ are unstable

for $\alpha \lesssim 1$, but the smallest damping rate for $\alpha \gtrsim 1$. Modes with $n \geq 1$ have monotonically decreasing oscillation frequencies while, for the fundamental mode, δ_I reaches a maximum value at $\alpha \approx 1.1$ and decreases afterward.

Critical α

For each mode n, a critical value of the cooling index, $\alpha_c^{(n)}$, may be defined, such that $\delta_R^{(n)} = 0$ when $\alpha = \alpha_c^{(n)}$. Hence, the nth mode is stable for $\alpha > \alpha_c^{(n)}$ and unstable when $\alpha < \alpha_c^{(n)}$, see Fig. 4. The value of the critical α is computed by interpolating α with a quartic polynomial passing through the two pairs of values across which δ_R changes sign. Thus the fundamental mode becomes stable for $\alpha > 0.388$, the first harmonic for $\alpha > 0.782$, and so on. Interestingly, the sequence of critical α is not monotonic with increasing n. Finally, notice that all (eight) modes become eventually stable for $\alpha \gtrsim 0.92$.

Linear Fit

By inspecting Fig. 3, one can easily see that, for a given α, the oscillation frequencies of the different modes are approximately equally spaced as n increases. In this respect, they resemble the modal frequencies of a pipe open at one end [21, 26, 32] and can be described by a simple linear fit of the form

$$\delta_I^{(n)} = \tilde{\delta}_I^{(0)} + n\Delta\tilde{\delta}_I, \tag{17}$$

with a small residual, $\lesssim 0.5\%$. In Eq. (17), $\tilde{\delta}_I^{(0)}$ is the "fitted" fundamental frequency and $\Delta\tilde{\delta}_I$ is a frequency spacing depending on the cooling index α. $\Delta\tilde{\delta}_I$ is monotonically decreasing for increasing α. Values of $\tilde{\delta}_I^{(0)}$ and $\Delta\tilde{\delta}_I$ are given in Table 1.

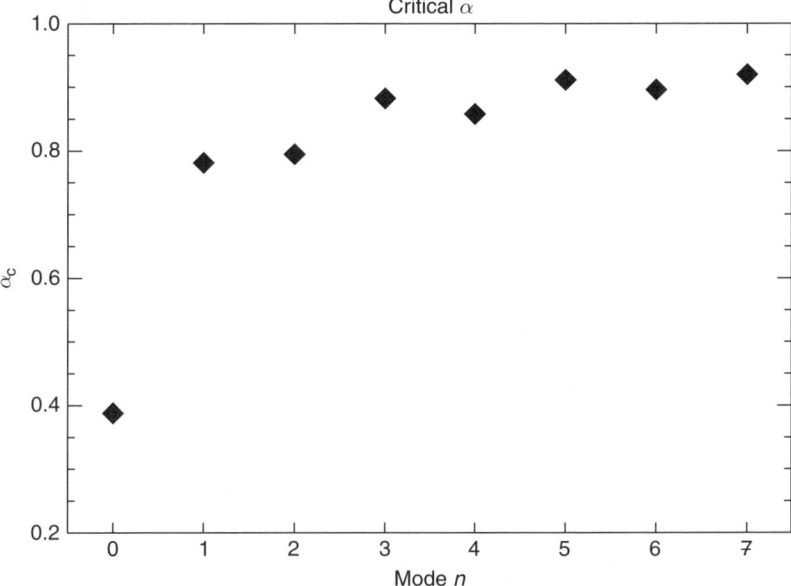

Fig. 4. Critical value of the cooling index as function of the mode number n. For a given mode n, values of $\alpha > \alpha_c$ have negative growth rates and thus are stable

Table 1. Fitted fundamental frequencies and spacing for the linear fit, Eq. (17)

	α							
	-2	$-3/2$	-1	$-1/2$	0	$1/2$	1	$3/2$
$\tilde{\delta}_{\mathrm{I}}^{(0)}$	0.2183	0.2352	0.2502	0.2642	0.2774	0.2898	0.3011	0.3076
$\Delta\tilde{\delta}_{\mathrm{I}}$	0.7486	0.7324	0.7129	0.6882	0.6553	0.6088	0.5359	0.3998

3 Nonlinear Dynamics

The results from Sect. 3 indicate that radiative shocks in real astrophysical settings may be linearly unstable and thus far from an equilibrium configuration. This calls for the investigation of the full time-dependent problem where nonlinear effects may play a major role in the shock dynamics.

In what follows, the radiative shock evolution is analyzed through a set of numerical simulations for different values of the cooling index α. We consider the shock in its initial steady-state, with the front being located at $x = 1$. For the sake of simplicity, flow quantities at the upper and the lower boundaries are held fixed to their initial values (see [19] for a comprehensive discussion on the choice of the boundary conditions). Numerical integration of Eqs. (1)–(3) is carried using the PLUTO code [20], a high-resolution Godunov-type code for astrophysical fluid dynamics. The onset of instability is triggered by the

discretization error of the numerical scheme and no external "ad hoc" perturbations are introduced, unless otherwise stated.

Issues concerning grid resolution effects must not be underestimated.
Sharp density gradients can be described with relatively limited accuracy because of numerical diffusion effects that cause high-density regions to "leak"
mass into neighboring low-density zones. Since the cooling process is proportional to the square of density, radiative losses will generally be overestimated,
causing abnormal, excessive cooling. Although this issue is intrinsic to any grid
of finite size and cannot be completely removed, higher resolutions can considerably mitigate the problem. Furthermore, small-amplitude oscillations of
the shock front can be adequately captured on finer grids.

3.1 Results

Figure 5 shows the density evolution in a time–space diagram for $\alpha = 0$.
During the early phases of evolution, the shock remains close to its equilibrium position, but around $t \lesssim 30$ small departures from the initial condition
begin to manifest. This stage is characterized by a linear growth of the perturbation: small density and pressure fluctuations tend to be enhanced as a
consequence of radiation effects, promoting faster cooling in regions of higher
density $(\Lambda \propto \rho^2)$. The lack of pressure support, in fact, amplifies strong non-
linear disturbances eventually steepening to form secondary shocks. These
secondary fronts propagate back and forth throughout the upstream region
and collide with the primary shock forcing the system into an oscillatory state.

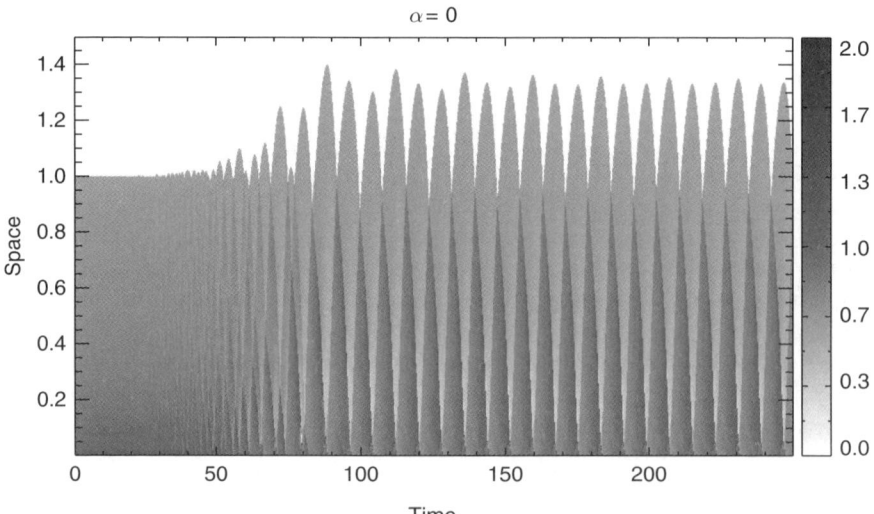

Fig. 5. Time–space diagram of the density logarithm for $\alpha = 0$. *Darker shades*
represent region of high density

For $t \gtrsim 80$, the amplitude of the oscillations begins to saturate and nonlinear effects become dominant by $t \approx 100$. During this phase, the largest oscillation peaks reach $\sim 40\%$ of the initial equilibrium position. For large t, the shock motion undergoes large amplitude oscillations upon which lower amplitude, high frequency modes superpose. In general, as shown in [19], the dominant mode of oscillation may be identified with the first overtone when $\alpha \lesssim 0.7$. Little power ($\lesssim 10\%$) resides in the fundamental mode.

The $\alpha = 0.5$ value (Fig. 6) is of particular astrophysical relevance, since it describes optically thin bremsstrahlung, which is the main source of radiative losses at temperatures of the order of 10^8–10^9 K, typical in accretion shocks in magnetic cataclysmic variables. Figure 6 shows that the solution remains close to the initial steady-state values for a longer time and unstable oscillations grow at a smaller rate. Indeed, when compared to the $\alpha = 0$ case, the oscillation amplitudes in the saturated regime are reduced by a factor of approximately 50%. It should be emphasized, however, that the transition to nonlinearity and the onset of the oscillatory cycle may depend crucially on the choice of the lower boundary condition, as observed in [19], especially when $0.4 \lesssim \alpha \lesssim 0.8$. The use of a cold dense gas layer (as opposed to the "fixed" boundary used here), for instance, acts as an absorber to incoming perturbations reducing the amplitude of the reflected waves. Even in presence of external "ad hoc" perturbations, the use of a cold dense layer may inhibit the growth of instability when $\alpha \gtrsim 0.45$.

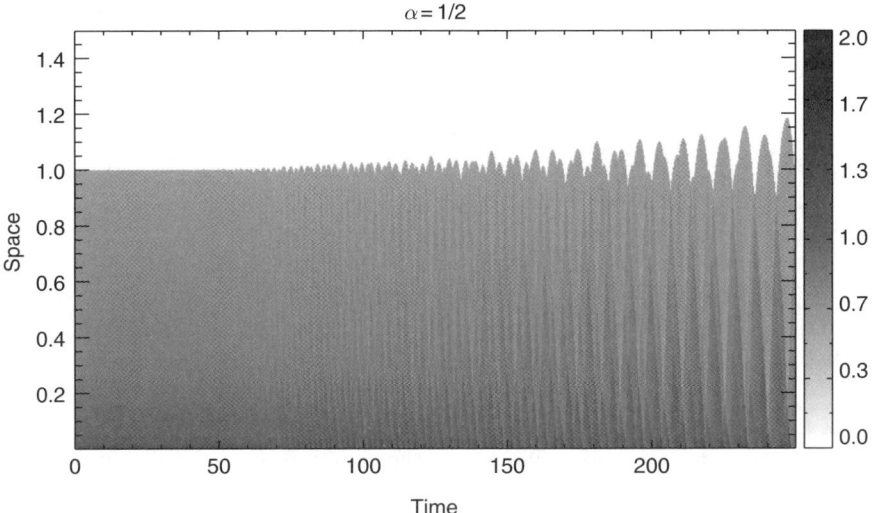

Fig. 6. Time–space diagram of the density logarithm for $\alpha = 0.5$. *Darker shades* represent region of high density

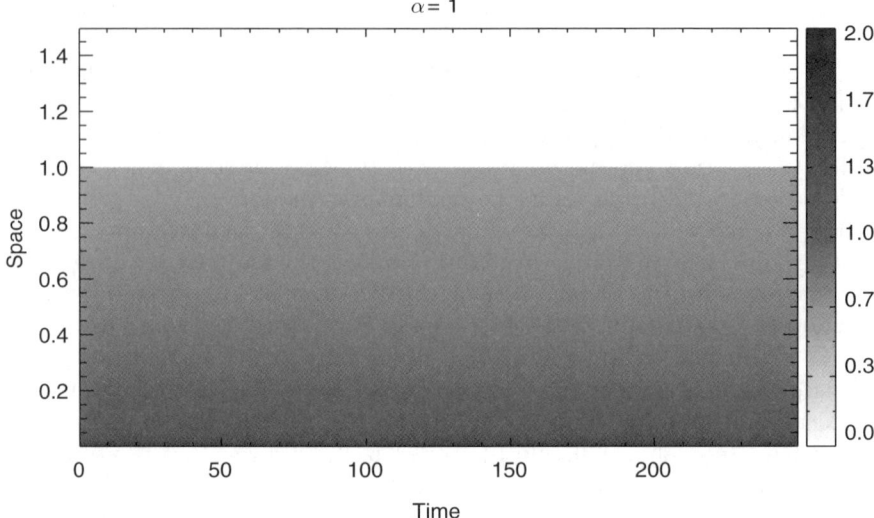

Fig. 7. Time–space diagram of the density logarithm for $\alpha = 1$. *Darker shades represent region of high density*

Finally, for $\alpha = 1$ (Fig. 7), grid perturbations are quenched and the system remains in equilibrium. This conclusion is confirmed even when external perturbation are introduced in the postshock flow (not shown here).

4 Discussion

A review of the instability of planar radiative shocks with a power-law cooling function $\Lambda \sim \rho^2 T^\alpha$ has been presented. Both linear stability properties and nonlinear time-dependent calculations have been reviewed.

Results from linear theory indicate that, for a given value of the cooling exponent α, multiple discrete modes of oscillation exist and the real and imaginary parts of the first eight eigenfrequencies have been derived. The overstable modes are labeled in the order of increasing oscillation frequency so that $n = 0$ corresponds to the fundamental mode, $n = 1$ to the first overtone, $n = 2$ to the second overtone, and so on. The stability criterion of a particular mode is expressed by the condition $\alpha > \alpha_c^{(n)}$, where $\alpha_c^{(n)}$ is the critical value of the cooling index for the nth mode. For the fundamental mode, for example, $\alpha_c^{(0)} = 0.388$, whereas for the first and second harmonic $\alpha_c^{(1)} = 0.782$ and $\alpha_c^{(2)} = 0.795$, respectively. A general trend towards stability exists for increasing α, so that all modes are stabilized for $\alpha \gtrsim 0.92$. It has been shown that oscillation frequencies are linearly proportional to the mode number n, a behavior similar to the quantized modes in a pipe.

The perturbative study has been complemented by several numerical simulations using the PLUTO code. The shock evolution has been followed through the linear and nonlinear phases for different values of α. These results agree with the perturbative study and show that the shock oscillations saturate at a finite amplitude and the system evolves in a quasi periodic cycle of collapse and reformation of the front. The amplitude of the oscillations can exceed $\sim 40\%$ of the shock equilibrium position and decrease for larger values of α. At $\alpha = 0.5$, the largest oscillations during the saturated phase are reduced to $\sim 20\%$ of the initial shock position. For $\alpha = 1$, the shock is stable and initial perturbations are damped on a characteristic time scale roughly proportional to the e-folding time of the first overtone. The late evolutionary phases reveal that the first overtone is the dominant mode of oscillation.

In spite of the oversimplifying assumptions adopted in this study, these results show a number of interesting consequences for a variety of astrophysical settings. Radiative shocks with velocities $v_s \gtrsim 130$ km s^{-1} are not uncommon in jets from young stellar objects, supernova remnants in the radiative phase, magnetospheric accretion in T-Tauri stars, and colliding stellar winds in relatively close binary systems. For these systems, the shocked interstellar gas reaches temperatures in the range 10^5–10^7 K and cools mainly by line emission, for which $\alpha < -0.5$. Under these conditions, radiative shocks are likely to show unstable behavior in all modes and phenomenological interpretations based on steady-state models become of questionable validity [13, 14]. Although inclusion of transverse magnetic fields extends the range of stability [26], the global thermal instability of radiative shock waves may still be important in interpreting a number of distinct observational features, such as emission-line ratios observed in interstellar radiative shocks [8], mixing between hot and cold material in colliding winds [1, 23], the filamentary structures observed in supernova remnants [3, 27], and so on.

Less conclusive assertions can be made for standing shocks in the accretion columns of Polar and Intermediate Polar systems. At temperatures of the order of 10^8–10^9 K, the X-ray emission is primarily determined by optically thin bremsstrahlung, although cyclotron and Compton cooling may not be neglected [21]. However, in the simple case where radiative losses are due to bremsstrahlung cooling only, $\alpha \approx 0.5$, the dynamics of the shock may be influenced by the interaction with the upper photospheric layers of the white dwarf [6]. Hence realistic models of accretion columns may require a more complex treatment of the lower boundary and additional physical processes.

References

1. Antokhin, I. I., Owocki, S. P., & Brown, J. C.: ApJ, **611**, 434 (2004)
2. Bertschinger, E.: ApJ, **304**, 154 (1986)
3. Blondin, J. M., Wright, E. B., Borkowski, K. J., & Reynolds, S. P.: ApJ, **500**, 342 (1998)

4. Chevalier, R. A. & Imamura, J. M.: ApJ, **261**, 543 (2003)
5. Calvet, N. & Gullbring, E.: ApJ, **509**, 802 (1998)
6. Cropper, M.: Space Sci. Rev., **54**, 195 (1990)
7. Falle, S. A. E. G.: MNRAS, **172**, 55 (1975)
8. Hartigan, P., Morse, J. A., & Raymond, J.: ApJ, **436**, 125 (1994)
9. Houck, J. C. & Chevalier, R. A.: ApJ, **395**, 592 (1992)
10. Imamura, J. N., Wolff, M. T., & Durisen, R. H.: ApJ, 276, 667 (1984)
11. Imamura, J. N., Rashed, H., & Wolff, M. T.: ApJ, **378**, 665 (1992)
12. Imamura, J. N., Aboasha, A., Wolff, M. T., & Wood, S. W.: ApJ, **458**, 327 (1996)
13. Innes, D. E., Gidding, J. R., & Falle, S. A. E. G.: MNRAS, **226**, 67 (1987)
14. Innes, D. E., Gidding, J. R., & Falle, S. A. E. G.: MNRAS, **227**, 1021 (1987)
15. Kimoto, P. A. & Chernoff, D. F.: ApJ, **485**, 274 (1997)
16. Langer, S. H., Chanmugam, G., & Shaviv, G.: ApJ, **245**, L23 (1981)
17. Langer, S. H., Chanmugam, G., & Shaviv, G.: ApJ, **258**, 289 (1982)
18. McCray, R., Kafatos, M., & Stein, R. F.: ApJ, **196**, 565 (1975)
19. Mignone, A.: ApJ, **626**, 373 (2005)
20. Mignone, A., Bodo, G., Massaglia, S., Matsakos, T., Tesileanu, O., Zanni, C., & Ferrari, A.: ApJ Supplement, **170**, 228 (2007)
21. Saxton, C. J., Wu, K., Pongracic, H., & Shaviv, G.: MNRAS, **299**, 862 (1998)
22. Saxton, C. J. & Wu, K. MNRAS, **324**, 659 (2001)
23. Stevens, I. R., Blondin, J. M., & Pollock, A. M. T.: ApJ, **386**, 265 (1992)
24. Strickland, R. & Blondin, J. M.: ApJ, **449**, 727 (1995)
25. Sutherland, R. S., Bicknell, G. V., & Dopita, M. A.: ApJ, **591**, 238 (2003)
26. Tóth, G. & Draine, B. T.: ApJ, **413**, 176 (1993)
27. Walder, R. & Folini, D.: A&A, **330**, L21 (1998)
28. Wolff, M. T., Gardner, J. H., & Wood, K. S.: ApJ, **346**, 833 (1989)
29. Wu, K., Chanmugam, G., & Shaviv, G.: ApJ, **397**, 232 (1992)
30. Wu, K., Pongracic, H., Chanmugam, G., & Shaviv, G.: PASA, Publ. Astron. Soc. Australia, **13**, 93 (1996)
31. Wu, K.: SSRv, Space Sci. Rev., **93**, 611 (2000)
32. Saxton, C. J.: Stability properties of radiative accretion shocks with multiple cooling processes. Ph.D. Thesis, University of Sydney (1999)

Index

ADER2 scheme, 56, 60, 63–66

Brio-Wu shock tube problem, 75

Cauchy-Kowalewski procedure, 20, 23, 24, 62
cooling index, 170
Courant number, 18

Euler Equations for general materials, 32
Euler equations, gas dynamics, 71, 166

flux limiter, SUPERBEE, 43, 47, 48, 51, 52
flux limiter, VANLEER, 48

general initial value problem, 4, 6–8, 10, 12
Godunov's theorem, 3, 26, 27, 43, 46

Haarten's theorem, 67
hyperbolic equations, 3, 16, 31, 60, 63, 66

Instability, Kelvin–Helmholtz, 106, 109, 154
Instability, Kelvin–Helmholtz, cylindrical geometry, 115, 123
Instability, Kelvin–Helmholtz, nonlinear evolution, 121
Instability, Kelvin–Helmholtz, physical properties, 118
Instability, Kelvin–Helmholtz, planar vortex sheet, 109

Instability, Kelvin–Helmholtz, radiative losses, 161
Instability, Kelvin–Helmholtz, relativistic flows, 119
Instability, Kelvin–Helmholtz, shear layer, 113, 121
Instability, Kelvin-Helmholtz, 132
Instability, macroinstabilities, 107
Instability, magnetic resonances, 135
Instability, magnetic shear, 135
Instability, MHD, 134, 137
Instability, microinstabilities, 107
Instability, oscillatory instability of a radiative shock wave, 164
Instability, Pressure-driven, 132, 134
Instability, thermal, 154
Instability, thermal, dispersion relation, 156
Instability, thermal, equilibrium condition, 154
Inviscid Burgers equation, 15
Isentropic Euler equations, 5

Kadomtsev's criteria, 144

Linear advection equation, 4–8, 11, 21, 26, 38, 46
Linear systems, 10

Monotone schemes, 43
MUSCL-Hancock scheme, 49, 56, 58, 59, 66

NUMERICA, 31

Numerical methods, finite difference method, 16, 17, 19, 28

Numerical methods, finite volume method, 28, 55

Numerical methods, forward in time central in space method, 19

Numerical methods, Godunov's method, 18

Numerical methods, Lax equivalent theorem, 20

Numerical methods, Lax–Friedrichs method, 19, 30, 45

Numerical methods, Lax–Friedrichs methods, 80

Numerical methods, Lax–Wendroff method, 20, 30, 38, 46

Numerical methods, Lax–Wendroff theorem, 31

Numerical methods, linear stability analysis, 24

Numerical methods, local truncation error, 21

Numerical methods, MHD equations, 72

Numerical methods, MHD, cell-centered methods, 92

Numerical methods, MHD, constrained transport, 95

Numerical methods, MHD, disontinu-ities, 76

Numerical methods, MHD, divergence cleaning, 93

Numerical methods, MHD, projection scheme, 94

Numerical methods, MHD, rarefaction waves, 77

Numerical methods, MHD, Roe scheme, 88

Numerical methods, modified equation, 24

Numerical methods, monotonicity, 25

Numerical methods, nonlinear, ENO, 43, 49, 53–55

Numerical methods, nonlinear, ENO, WENO, 54

Numerical methods, nonlinear, TVD, 43, 45, 46, 49, 51, 52, 55, 58

PLUTO, 75, 171, 175

Powell's 8-wave formulation, 92

Rankine-Hugoniot jump conditions, 75, 166

Riemann problem, 13, 72

Riemann problem, MHD, 73

Riemann solver, 33, 41, 59

Riemann solver, ALICE, 36

Riemann solver, EVILIN, 3, 32, 37

Riemann solver, HLL, 41, 80

Riemann solver, HLLC, 3, 32, 82

Riemann solver, HLLD, 85

Riemann solver, linearized, 38

Riemann solver, MUSTA, 37

Riemann solver, Rusanov solver, 80

Riemann solver, two-rarefaction, 36

Riemann solver, two-shock, 37

shock front, 166

shock wave, 165, 166

Sommerfeld conditions, 111

Suydam's criterion, 134, 135

Sweby TVD region, 46, 47

The Energy Principle, 140

Traffic flow equation, 5

WAF scheme, 64–66

Wave speed estimates, 34, 36, 37